Materials and Processing Failures in the Electronics and Computer Industry: Analysis and Prevention

A.S. Brar, P.E.

and

Prativadi B. Narayan, Ph.D., P.E.

Acquisitions Editor - Mary Thomas Haddad
Production Project Manager - Suzanne E. Hampson
Production/Design - Randall L. Boring

The Materials
Information Society

First printing, October 1993

This book is a collective effort involving hundreds of technical specialists. It brings together a wealth of information from worldwide sources to help scientists, engineers, and technicians solve current and long-range problems.

Great care is taken in the compilation and production of this Volume, but it should be made clear that NO WARRANTIES, EXPRESS OR IMPLIED, INCLUDING, WITHOUT LIMITATION, WARRANTIES OF MERCHANTABILITY OR FITNESS FOR A PARTICULAR PURPOSE, ARE GIVEN IN CONNECTION WITH THIS PUBLICATION. Although this information is believed to be accurate by ASM, ASM cannot guarantee that favorable results will be obtained from the use of this publication alone. This publication is intended for use by persons having technical skill, at their sole discretion and risk. Since the conditions of product or material use are outside of ASM's control, ASM assumes no liability or obligation in connection with any use of this information. No claim of any kind, whether as to products or information in this publication, and whether or not based on negligence, shall be greater in amount than the purchase price of this product or publication in respect of which damages are claimed. THE REMEDY HEREBY PROVIDED SHALL BE THE EXCLUSIVE AND SOLE REMEDY OF BUYER, AND IN NO EVENT SHALL EITHER PARTY BE LIABLE FOR SPECIAL, INDIRECT OR CONSEQUENTIAL DAMAGES WHETHER OR NOT CAUSED BY OR RESULTING FROM THE NEGLIGENCE OF SUCH PARTY. As with any material, evaluation of the material under end-use conditions prior to specification is essential. Therefore, specific testing under actual conditions is recommended.

Nothing contained in this book shall be construed as a grant of any right of manufacture, sale, use, or reproduction, in connection with any method, process, apparatus, product, composition, or system, whether or not covered by letters patent, copyright, or trademark, and nothing contained in this book shall be construed as a defense against any alleged infringement of letters patent, copyright, or trademark, or as a defense against liability for such infringement.

Comments, criticisms, and suggestions are invited, and should be forwarded to ASM International.

Library of Congress Cataloging-in-Publication Data

Narayan, P.B. (Prativadi B.)
    Materials and processing failures in the electronics and
        computer industries: anaylysis and prevention/P.B. Narayan
        and A.S. Brar
            p.   cm.
    Includes index.
    ISBN 0-87170-468-4
    1. Electronics—Materials—Defects.
    2. Electronic apparatus and appliances—Defects.
    3. System faulures (Engineering)
    4. Electronic industries—Case studies.
    I. Brar, A.S. II. Title.
    TK7871.N37   1993
    621.381—dc20   93-5530
                                         CIP

                ISBN: 0-87170-468-4

            ASM International®
        Materials Park, OH 44073-0002

        Printed in the United States of America

Acknowledgements

A.S. Brar thanks his colleagues at Control Data Corporation who helped solve many of the failure problems discussed in the case studies included in this volume. He feels it is difficult to remember everyone to whom thanks is due, and to acknowledge every book and every piece of technical information that contributed to his knowledge and personal growth. He says his best teacher has been the materials and process failures themselves. He acknowledges the encouragement he received from his wife and children. He is gratified that readers may benefit through this book from his practical experience in problem-solving, and based on this may increase the quality of the manufactured products with which they are involved.

Prativadi B. Narayan thanks the engineers and management of Control Data Corporation, StorageTek, and Rocky Mountain Magnetics, companies where he has worked, for their encouragement and support in his endeavors to understand the fundamentals of materials processing. He admires and respects the American management system that stimulates innovation and intellectual growth in professional employees. He is grateful to his father, Dr. P.B. Acharya for cultivating in him love for "the magic of science and technology." He gratefully acknowledges the encouragement and realization of the importance of education he received early in life from his mother, Mrs. Swarajyalakshmi; brothers, Ranganath and Srinivas; sister, Indira; aunt Mrs. Suguna; and uncle, Mr. Narasimhacharya. He appreciates the cooperation and encouragement of his wife, Indira; sons, Sreenath and Madhav; and daughter Suguna shwo made writing this book a great fun. Their affection and enthusiasm has been an inspiration for each and every step.

The authors would like to thank Drs. Ram Natesh, K.L. Mittal, and J.P. Sharma for reviewing the manuscript and providing valuable feedback. They would like to acknowledge the permissions of editors of the following publications, along with their publishers, for allowing them to reprint previously published material: *Computer Technology Review, Connection Technology, Materials Engineering, Microcontamination, Microelectronic Manufacturing and Testing, Products Finishing, Research and Development, Test & Manufacture World,* and San

Francisco Press, Inc. for allowing them to reprint material from conference proceedings.

Both authors thank Mrs. Mary Thomas Haddad and the rest of the staff at ASM International for their courtesy, patience, diligence, and professional excellence.

Preface

Manufacturing is the key to survival of industrial economies and sustenance of comfortable living standards. The manufacturing process is a series of materials processing operations, and it is necessary to fully characterize each one to obtain a robust process flow. Because reliability is particularly important in the electronics industry, reliability of the product must be constantly improved by identifying and optimizing the processing operations that have the greatest impact on it. Process failure analysis identifies the weakest link in the manufacturing process chain and determines where technical and financial resources should be directed for maximum efficiency.

In designing and manufacturing a product, it is essential to know and understand as many different modes of failure as possible so that the concept of reliability is introduced even at the design stage. The most efficient way to produce a zero-defect part is to avoid all known modes of failure and eliminate those that appear during manufacturing, testing, and use.

The cost of failure of a component increases exponentially, depending on how late in the process flow the failure is detected. For example, a thin-film read/write head failure costs a few cents when detected at the wafer processing stage, a few dollars when detected at the slider stage, a few hundred dollars in a disk drive inside the manufacturer's company, and several thousand dollars at the customer's site (excluding customer good will, which cannot be easily measured in dollars and cents). In addition to marketing and other advantages, it is cost-effective to detect and rectify modes of failure at the earliest stage of manufacturing.

To characterize a failure, the process flow should be examined step by step to determine exactly where the failure appeared and where the process deviated from the optimum. Based on the experimental evidence, a cause of failure is hypothesized. A thought experiment is conducted according to the process flow, and the results of that experiment are compared to the actual evidence, which confirms or denies the original hypothesis. If the hypothesis is denied, the next best hypothesis goes the same route until its consequences (or those of a subsequent one) match the experimental results.

The primary objective of failure analysis is to find out how to avoid every mode of failure. To stress the preventative aspect of failure analysis, discussion of each case study in this volume is followed by recommendations and a discussion of their effects. In a sense, process failures indicate potential problems the customer could face in using the product. As a result, failures should be regarded as prized evidence of the Achilles heel of the manufacturing process flow and should be taken care of expeditiously. A failure is similar to a customer complaint about a product or service. So few customers actually take time to file a complaint that each one who does provides a valuable glimpse of the actual experiences of the customers and should be resolved with the highest priority.

This volume contains numerous case studies collected in the authors' 35 years of combined experience in materials failure analysis, process improvement, and reliability enhancement. Although many of the examples come from working in the magnetic recording and computer industry, a process failure belongs to the deficiency of that process and is relevant to any industry that uses it.

No two failures are identical in every respect. However, experience and familiarity will help in focusing on the important aspects of the problem, thus leading the way to a quicker and more intelligent solution.

Because of component miniaturization, microcontamination poses the greatest threat to reliability in the computer and electronics industries. Each component is evaluated based on its functional reliability and capacity to restrict generation of microcontamination. For example, when a magnet generates particles, there may not be any noticeable decrease in the magnetic field it produces (i.e., it may continue to be functionally reliable), but the loose contamination could destroy the product. This volume contains an extended discussion of sources of microcontamination, how it is generated, and how it can be eliminated by using protective coatings, surface cleaning techniques, design modifications, and other methods. The many case studies of failures caused by microcontamination underscore the importance of maintaining clean rooms, adhering to good clean-room practices, testing every component as a potential source of contamination, and establishing a process flow that produces minimal or no contamination.

The authors are confident this volume will be helpful to scientists, engineers, and technicians who have an interest in materials and processes and who work in design, development, manufacturing, quality control, and reliability groups in the computer and electronics industries, and to vendors who provide components and services such as joining, protective coatings, and plating.

The authors hope that the knowledge gained by perusing the case studies will enable the readers to anticipate materials and processes failures and thereby realize zero-defect products. If every reader of this volume can eliminate at least one failure in his or her workplace, the authors will be gratified that their endeavors have paid a handsome dividend.

P.B. Narayan
A.S. Brar

Table of Contents

Chapter 1

Introduction to the Philosophy of Failure Analysis

1.1 Traditional Concept of Failure Analysis

In the traditional sense, failure analysis implies the occurrence of a total failure, such as a mechanical fracture or an open circuit, followed by analysis to determine the cause of that failure. The traditional main objective of failure analysis was to catalog the failure in a broad sense and issue a report concerning all aspects of it, aimed at permitting correction of the problem and the development of measures to prevent its recurrence. This information was given to the product development or manufacturing divisions for their interpretation and further action. In the past, the mission of the failure analysis organization was simply to analyze failures of the end products, i.e., find out how they occurred. It was not encouraged to investigate which part of the process was responsible for the failure, because failure analysis was not regarded as an integral part of the development and manufacturing process.

1.2 Modern Definition of Failure Analysis

But the realization that, without knowledge of the source of the problem, the failure mode cannot be eliminated, has changed this. Failure analysis is now defined much differently: its main objective is failure *prevention*.

How can failure be prevented in a product? By ensuring that all of its components meet their respective specifications. Any component that, either in product development or manufacturing, fails to meet specifications, should be termed a failure for failure analysis purposes. Whether a resistor with a specified resistance of $100 \pm 10\Omega$ exhibits a resistance of 80Ω, or a magnetic medium with a specified magnetic coercivity of 1000 ± 50 Oe exhibits a coercivity of 900 Oe, or a tribological combination (e.g., ball bearing or magnetic head-disk interface in a Winchester drive) with a service life of at least 20,000 contact-start-stop cycles shows unacceptable wear or friction buildup in 15,000 cycles, these conditions constitute failures that are candidates for failure analysis.

Modern manufacturing theories, spearheaded by Taguchi (Ref 1) and Kacker (Ref 2), extol the need for perpetually tightening specification limits on the properties of all components in a given part, to improve overall quality. As such, there is an increasing need for failure analysis to assist in improving the quality of any product.

1.3 Critical Features of the Microelectronics Industry

In the semiconductor industry, there is an ever-increasing need to accommodate more components on the same chip, even though line widths are already in the submicrometer range. In computer hard disk drives, the magnetic medium on modern metallic disks is 30 nm (1.2 μin.) thick with a 15 nm (0.6 μin.) thick overcoat. The read/write head is positioned ("flies") 100 nm (4 μin.) above the disk. The magnetoresistive head technology uses metallic thin films with thicknesses of only 30 nm (1.2 μin.). This need for miniaturization presents three important problems from a materials point of view:

First, it imposes stringent restrictions on the material properties of the components, making the specification limits very narrow. Control of metallurgical features, such as grain size, second-phase particle distribution, and composition, is essential for the materials used.

Second, it necessitates widespread use of multilayers of thin films only several nanometers thick. These thin films are quite fragile, and because of their thinness their surface properties have a significant effect on their integrity. Material degradation caused by corrosion and solid-state diffusion becomes a crucial factor in component durability. Material degradation processes often have an exponential temperature dependence. Humidity and chemically active species such as chlorine strongly affect corrosion and wear. Minor variations in environmental factors, such as temperature, humidity, and the presence of undesirable species (e.g., chlorine and sulfur), can have a disastrous effect on the durability and reliability of the component, and these variations must be kept under strict control.

Third, microcontamination is the bane of the microelectronics industry. In semiconductor materials it causes electrical short-circuiting. In computer disk drives it causes wear on the magnetic medium, which can lead to a catastrophic loss of information, called a "head crash." It can also increase static friction (stiction) at the head-disk interface, which can render the drive unusable.

Microcontamination can be classified in two categories. One is particulate contamination, which includes atmospheric dust, burrs, and shavings on the metal parts after machining; wear debris; solid corrosion products; salts; and organic and metallic fibers. The other is film-like hydrocarbon contamination, which is more insidious than particulate contamination, primarily because of the difficulty of detecting organic films. The sources of film-like contamination are machining oils, human handling, perfumes, gaskets, resins, and plastic parts.

Outgassing of plastic materials over a period of time makes a significant contribution to microcontamination. Thermal analysis is a useful tool to monitor potential contamination problems in plastics (see Appendix 1.1).

The discussion of failure and failure analysis in the following chapters will concentrate on failures, their analysis and prevention, in light of the special requirements of the microelectronics industry, such as tight tolerances, miniature sizes, and microcontamination. It will stress the goal of reducing possible sources of microcontamination. While it is often difficult to establish whether or not a particular contaminant caused or contributed to a failure, it should always be assumed that any contamination is detrimental to continuing service life, and wherever possible it should be eliminated at its source.

Microelectronics manufacturers expend substantial amounts of time and resources to control contamination in manufacturing areas. For many applications, manufacturing is conducted in Class 10 clean rooms. Maintaining the requisite air flow for these facilities, ensuring its proper filtration, and providing for constant removal of contaminants from manufacturing personnel in the clean rooms, all constitute an enormous task. In this context the microcleanliness requirements that must be placed on materials and coatings used in making these components can readily be appreciated.

1.4 Causes of Failures

In general, failures occur because of improper selection of materials, lack of control in processing, and unexpected environmental conditions. In addition, a combination of these three conditions leading to failure could be encountered in many manufacturing circumstances.

1.4.1 *Materials Selection*

Proper selection of the material for a particular component requires a clear understanding of how that component is actually used in service. For example, what thermal, mechanical, and electrical stresses will it be required to withstand? With what materials will it interact?

Quality and reliability are the first and foremost criteria to be considered in materials selection, followed by manufacturability, availability, and cost. For example, it is better to select an alloy that is widely used, even if it is slightly more costly, than to select an obscure alloy based primarily on cost considerations. Materials vendors have more information about, as well as experience with, well-known alloys. A thorough understanding of the operation of the complete product and the role of the specific component in that product is necessary before an informed materials selection decision can be made.

In the case of metals, information on what materials are in electrical contact with one another is necessary, because of the possibility of galvanic corrosion and consequent material degradation. When two dissimilar metals are in contact with one another in the presence of an electrolyte, the metal that is electrochemically more negative becomes corroded, while the electropositive metal is protected from corrosion.

When two dissimilar metals must be connected, they should be as close as possible to each other in the electrochemical series. The farther apart they are electrochemically, the greater the potential for corrosion. The more noble metal should occupy the smallest area possible. Because the presence of an electrolyte is essential for galvanic corrosion to occur, the amount of it that accumulates at the metal junction should be minimized. Also, the use of corrosion-resistant and passive coatings reduces galvanic action.

Organic coatings and thin films, both metallic and ceramic, play an important role in tailoring surface properties, such as corrosion, reflectivity, and tribology, to the specific application. Powder coatings and electrostatic spray coatings significantly improve corrosion resistance, although they do not provide an adequate barrier against moisture penetration. Most of these coatings are electrical insulators. Consequently, when a part needs to be machined for electrical grounding, the interface of the machined and coated areas is a potential source of particle generation. These coatings tend to outgas when used in high vacuum and ultra clean environments. However, their chemical inertness and economy of application render them very attractive for many noncritical applications.

Chromate conversion coatings provide reasonable corrosion resistance on alloys of aluminum, magnesium, and iron (plain carbon or low-alloy steel). These coatings are affected by repeated washings in deionized water and by exposure to temperatures above 80 °C (175 °F) and consequently may not meet the stringent cleanliness requirements of critical applications, particularly on cast alloys. With aqueous cleaning becoming the primary alternative to chlorofluorocarbon cleaning, the use of chromate conversion coatings is decreasing rapidly.

Metallic coatings such as electroless nickel and ion vapor deposited coatings are applied by plating and vacuum techniques such as sputtering, chemical vapor deposition, and thermal evaporation. Many of these coatings are discussed in later chapters. For a brief introduction to coatings, please see Appendix 1.2.

1.4.2 *Process Control*

Lack of control in processing materials is the second most important cause of failure. Process control is currently receiving much attention as a means of increasing quality and productivity, and justifiably so. The use of statistical process and quality control is becoming popular in the manufacturing sector.

Materials must undergo processing to attain the structural properties required for service. For example, consider the sequential sputter deposition of thin films to obtain a multilayered structure. Sputter deposition process variables include argon gas pressure, background gas pressure, voltage, current, power, and substrate temperature. The effect of process variables on the materials properties of the component and on the performance characteristics of the final product needs to be studied.

A statistical experiment based on Taguchi experimental design, for example, provides an understanding of which parameters are critical, with the fewest number of experimental runs. A thorough understanding of the effect of the combination of variables on materials properties will help identify the most important variables to be monitored.

Let us consider another example: heat treatment. In a material such as steel, resultant mechanical properties closely depend on the conditions of heat treatment, such as surface cleanliness of the material, heating rate, temperature of heat treatment, environmental composition, cooling rate, quenching medium, etc. The design of a process should be such that the properties of the product are relatively insensitive to minor fluctuations in process variables.

In manufacturing, it is expensive to maintain a process variable such as temperature under very close control. Instead, it is more economical to allow for more liberal process variations, provided the proper material is selected or the process is designed so that the material properties are relatively insensitive to process fluctuations.

1.4.3 *Environmental Effects Including Microcontamination*

The third most common reason for failure is an undesirable change in the environment, in such factors as ambient temperature, humidity, and the arrival of corrosive species (e.g., chlorine), microcontamination generated by other components, and handling contamination. For example, most modern computer disk drives are sealed to keep out atmospheric contamination and provide a clean and controlled operating environment inside the drive. If their design incorporates breathing filters to equalize inside and outside atmospheric pressure, then humidity and corrosive species can slowly migrate into the drive.

High ambient temperatures, in general, increase the rate of corrosive reactions. They also affect the tribological properties of the head-disk interface in a disk drive. The product faces many more reliability challenges when used in a shop floor environment as opposed to a computer room, where temperature and humidity are controlled throughout the year. For example, if the product is used in a plant that manufactures hydrochloric acid near a sea coast, both humidity and chlorine concentrations there will be significantly higher. Since many metals degrade in such environments, the reliability of the product could be severely challenged. The product must be designed for the most challenging environment and must pass the worst case scenario qualification tests.

If any of the components enclosed in a clean room or clean environment within a product (e.g., a disk drive) generates microcontamination of either particulate or film-like nature, the reliability of the manufacturing process or the product will be seriously jeopardized. Each and every component used in the product must be tested to ensure that it is not a source of microcontamination.

Total robotic manufacturing is still not a reality. As long as human beings are involved in the manufacturing process, there is a chance of human or handling contamination. Good clean-room practices are essential in manufacturing areas to eliminate contamination such as spittle marks, perspiration, body oils, hair splinters, fragrances, and cloth fibers.

1.5 Methodology of Failure Analysis

Before a failure can be analyzed, as much information as possible should be obtained about the product in which the failure event occurred. Important considerations include:

- Does the part contain reconditioned or reworked components?
- What type of environment was it exposed to before the failure occurred?
- Are any neighboring parts exhibiting abnormal behavior?
- Where do the materials properties of each component fall on the normal distribution curve for that particular component?
- Do any of the components come from a suspect manufacturing lot?

What symptoms of malfunctioning associated with the failure were noticed? Was there any unusual noise? Is there any indication of unexpected heat generation or a sudden increase in vibration? Are there increased contamination levels (e.g., an increase in particle count, if a particle counter is being used) or an increase in the error count?

To begin the physical analysis, the failed product should be opened in a clean room. Anything unusual about the appearance of any component within it should be noted. Visual examination followed by optical microscopy studies will indicate which areas need to be examined with a scanning electron microscope to elucidate morphological details and which areas need to undergo energy-dispersive X-ray spectroscopy to determine compositional analysis. These analytical techniques and their capabilities and limitations are discussed in Chapter 2.

The following questions should be addressed in the failure analysis report:

- How did the failure manifest itself? What was the first indication of failure?
- How was the failure analyzed? What analytical instrumentation was used for the analysis?
- What test data were used to reach the conclusion arrived at?
- What corrective action is suggested for eliminating this type of failure? What are the results of the corrective action? Did it solve the problem?

Information concerning the failure should be organized in such a manner that it will be readily available without having to repeat the analysis if a similar failure occurs in the future. This approach saves time, money, and effort.

1.6 Challenges of Failure Analysis

In analyzing failures, the sequence of events that led to the failure must be determined. For example, the contaminant particle could be changed in size, shape, texture, color, and chemical composition through action of the various processes leading to the failure. The evidence at the failure site shows the contaminant as it appeared during or after the failure, and it cannot show its original condition or the sequence of changes it experienced. Based on the evidence, a hypothesis of how the failure occurred should be formulated and then compared with the analysis data to check its veracity. Particular attention must be paid to any experimental findings that contradict the original hypothesis, which should be modified to fit the contradicting data. The more iterations and modifications that occur, the closer the hypothesis will come to the actual failure mechanism.

It is important to understand how each component is affected by the presence of the other components and the environmental changes. Also, the interplay of various components can complicate the way failure appears. Failure characterization cannot be effectively made from an armchair analysis. Persons responsible for it must be closely in touch with the details of the manufacturing process and product function.

Electrical, mechanical, physical, chemical, and material properties should be taken into consideration before the final cause of failure is determined. The manner in which these variables interact and their relationships with one another are also important considerations to address before a final decision is rendered.

Failure analysis is the meeting ground for the physical sciences and engineering disciplines, and an understanding of the interrelationship between them is necessary for effective analysis. The intuitional ability to discern which aspect of a material/service condition is dominant in a given failure mode is valuable. For example, a case study of a bearing failure will be discussed in the later chapters, in which poor lubricant conductivity caused an electrical charge to build up between the two rails. This led to an electrical discharge that heated the local areas, melting the metal and generating particulate contamination of the metal and oxide. As temperature increases, the rate of oxidation increases almost exponentially. Oxides are hard and abrasive and can damage the bearing surface. Solving this failure problem involved understanding the basic principles of physics, chemistry, mechanical engineering, and materials engineering.

1.7 Failure Prevention: The Essence of Failure Analysis

Let us recount an incident involving Dr. Robert Oppenheimer, the eminent physicist who was the leader of the scientific group that developed the nuclear bomb at Los Alamos Laboratories in the 1940s. After the first successful nuclear test, the scientific community gained an understanding of the enormous destructive capability of the chain reaction. Dr. Oppenheimer addressed a U.S. Congressional committee to explain the full ramifications of the nuclear bomb.

The Congressmen were justifiably horrified by the enormity of the consequences of the explosion of the nuclear bomb and asked him, "Is there any safeguard you can think of against the horrible nuclear explosion?"

Dr. Oppenheimer replied, "Sure, I can."

"And that is...?", asked a curious Congressman.

There was a hush of silence and expectation.

Dr. Oppenheimer calmly replied, "Peace."

This story illustrates the underlying philosophy of studying failure for the purpose of improving quality. What is the main function of failure analysis? How should industry make sure that a product operates without the danger of failure? What should be done to protect it from failure?

The answer to all these questions is to ensure that failure does not occur in the first place. It is very expensive to take remedial action once failure has occurred. Each example of failure analysis should provide another means of failure avoidance. Failure analysis is similar to a physical examination by a physician; the main objective is to determine potential problems that could adversely affect the health of the person. The examination should be as nondestructive as possible. Failure analysis should not be relegated to the status of post-mortem examination; it should be an annual physical examination.

References

1. (a) G. Taguchi, *Experimental Design*, 3rd ed., Maruzen Publishing Company, Tokyo, 1976, in Japanese; (b) G. Taguchi and Y. Wu, *Introduction to Off-Line Quality Control*, Central Japan Quality Control Association, 1980; available from American Supplier Institute, 32100 Detroit Industrial Expressway, Romulus, MI 48174

2. R. Kacker, Off-Line Quality Control, Parameter Design, and the Taguchi Method, *J. Quality Technol.*, 17(4), 176, 1985

Appendix 1.1

Materials Evaluation of Disk Drive
Components by Thermal Analysis*

High-density disk drives are expected to provide up to 30,000 bits/in. (bpi) and 2,000 tracks/in. (tpi). To do this, the read/write heads must fly at heights of less than 0.25 μm above the hard disk. These requirements demand very tight dimensional tolerances on many disk drive components and, in turn, require that component parts meet very precise materials specifications.

Components used in drives should neither generate particles nor outgas. Particulates, as well as contaminant films (mainly hydrocarbons), are known to interfere with head flight, cause magnetic media wearout, increase static friction (stiction), and ultimately lead to head crash and catastrophic loss of information.

New materials, especially organic substances, are constantly being developed and tested for use in disk drives. In general, the quantity of liquids and volatile species evolved from them should be minimized to avoid contaminant migration throughout the disk drive environment. Because organic materials are quite sensitive to changes in temperature, thermal analysis techniques are found to be very helpful in determining materials characteristics such as weight changes, heats and temperatures of transitions, dimensional changes, and mechanical damping and modulus characteristics in specified environments.

---

* P.B. Narayan, A.S. Brar, and Y.M. Chow, *Solid-State Technol.*, May, 1989, pp. 250-253. Reprinted with permission.

Thermal Analysis

Thermal analysis of a material involves measuring certain of its physical properties as a direct or indirect function of temperature and heating rate. Four of the most commonly used thermal analysis techniques are thermogravimetric analysis (TGA), differential scanning calorimetry (DSC), dynamic mechanical analysis (DMA), and thermomechanical analysis (TMA). Each provides a unique piece of information.

If an organic material undergoes a weight change in passing through a transition temperature (due for example to loss of one of the components by outgassing, reaction with the environment, etc.), then TGA could be used to monitor the transitional kinetics. On the other hand, if heat flow occurs due to an exothermic or endothermic reaction, then DSC may be used to study the change at the transition temperature. In situations where there is no weight change or heat flow, DMA is appropriate for investigating mechanical damping and modulus response characteristics and for keeping track of the transitional kinetics of a process. In still other cases involving thermally induced dimensional changes, TMA can be used to study the me-

Table 1   Thermal analysis for disk drive components

Technique	Measurements	Applications
DSC: measures heat and temperature of transitions	Specific heat, heat of transition, degree of cure, kinetics of oxidation, and thermal stability	Cure and processing properties of coating resins; thermal characteristics of photo-resists; cure of adhesives; stability of polymer material used in drives
DMA: measures mechanical properties (modulus and damping)	Glass transition, secondary transition temperature, tensile and shear modulus	Phase separation of coating resins; cure advancement of resin system
TGA: measures weight changes	Decomposition temperature, thermal stability, compositional analysis	Amount of outgassing; resin/filler ratio; decomposition kinetics of lubricants; thermal stability of gaskets, adhesives, etc.
TMA: measures dimensional changes	Expansion/contraction, softening temperature	Metal expansion; substrate/coating compatibility

chanical responses or to determine the softening temperature of materials (Table 1).

Thermal analysis effectively determines the temperature and temperature/time characteristics of a process. Property changes as a function of temperature, heating rate, and time at temperature can be monitored either in air (mimicking the environment in which most products function) or in an inert gas (e.g., nitrogen) environment to develop a clearer understanding of a specific thermal response.

Heating rate is analogous to the strain rate during a tensile strength measurement. It is an important parameter because, when it is very fast, the thermal response of a specimen may lag and measurements may indicate a higher transition temperature, or occasionally miss an intermediate transition altogether. Therefore, a thorough understanding of a material is needed before a suitable heating rate is chosen. It is general practice to mention the heating rate for all thermal analysis results.

Thermogravimetric Analysis

Thermogravimetric analysis (TGA) measures the weight gain or loss of a material as a function of temperature and/or time in a specified environment. It is widely used to study outgassing, decomposition, and the thermal stability of plastics, adhesives, gaskets, rubber, lubricants (Fig. 1), and other organic materials, but it may also be used in weight change studies aimed at choosing the correct resin/filler ratio for a material and optimizing the curing cycle. Organic materials can be tested for longevity and reliability, thermal degradation or degradation due to environment (e.g., oxygen, water, etc.), weight change to determine curing conditions of resin based composites, and adsorption or absorption of water or solvents during parts cleaning cycles. However, in the latter case, care needs to be taken in assigning weight loss to absorption because a material could outgas, making it difficult to separate the two phenomena.

The analysis is important in disk drive component analysis because any plasticizer or hydrocarbon that evolves from organic materials during the life of a drive could collect at the head/disk interface and interfere with head flight. In addition, such loss of a volatile component could also degrade the materials properties of the plastic part.

Thermogravimetric analysis is performed while a specified gaseous environment is maintained around the specimen. Disk drives are

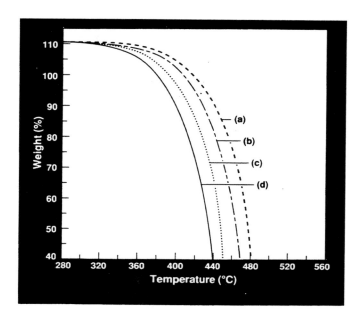

Fig. 1 TGA thermogram of a lubricant. Lubricant lifetime can be estimated from the weight loss measurements. Heating rates are (a) 20 °C/min; (b) 10 °C/min; (c) 5 °C/min; and (d) 2 °C/min.

designed to work properly at temperatures of 10 to 65 °C (50 to 150 °F) and at a relative humidity of 20 to 80%. One wants to determine whether a particular organic can survive in the disk drive atmosphere at a given temperature and relative humidity. Winchester drives, for example, are generally sealed, but different drives allow for varying amounts of "breather" or exchange of air with the outside atmosphere. Because the mean time between failures (MTBF) of a drive should be several years, accelerated reliability testing is important.

Thermogravimetric analysis equipment may also be used to determine the Curie temperature of magnetic materials, i.e., the temperature point at which a ferromagnetic material loses its magnetism and becomes paramagnetic. Because ferromagnetic materials are present not only in the disk storage medium but also in various disk drive components (e.g., the spindle motor), TGA also presents itself as a readily available magnetic materials quality control procedure.

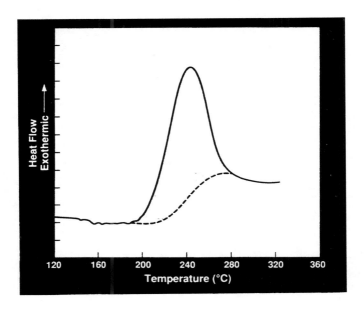

Fig. 2 DSC cure curve of a thermoset material. Analysis of the heat flow versus temperature is used to optimize the cure process. The hearing rate is 10 °C/min, and the heat of reaction (the shaded area under the curve) is 325 J/g.

Differential Scanning Calorimetry

With differential scanning calorimetry (DSC), one measures heats and temperatures of transitions. A specimen is placed in a furnace and heated at a specified rate. If an exothermic or endothermic reaction occurs, net heat flow, or the energy of the reaction, is measured by the DSC instrument. Crystallization, oxidation, and some decomposition reactions are exothermic, whereas phase transitions, reduction, dehydration, and other decomposition reactions are endothermic.

Differential scanning calorimetry has been found to be very useful in optimizing the curing procedure for the resins used in drives. By plotting the heat flow as a function of temperature, the curing temperature can be measured (Fig. 2). Alternatively, by measuring the heat flow as a function of time, at a constant temperature (isothermal plots), the curing time may be determined. Many of the disks available at present contain particulate iron oxide magnetic media, in which the oxide particles are bound by an epoxy resin. The resin should be completely cured so that the medium attains its specified toughness. If curing is incomplete, the medium could wear out easily during its interaction with the head.

During manufacture of an oxide disk, iron oxide (with a small percentage of alumina added for wear resistance) is milled with epoxy resin, solvent, and surfactants. The magnetic paint thus obtained is coated on an aluminum substrate. After orienting the oxide particles, the disk is baked to cure the resin and polished to obtain the required surface finish. Differential scanning calorimetry is used to optimize both the temperature and time of paint curing. Curing parameters are periodically checked because media toughness, which depends on curing, is the most important parameter for reliable head/disk operation.

Many fragile parts of a disk drive are bonded with organic adhesives. Strict dimensional tolerances and other restrictions require that only small amounts of resin be used. Complete curing is, therefore, essential for developing proper bonding strength. Here, too, DSC finds application.

Dynamic Mechanical Analysis

If a resin system does not show weight change or measurable heat of reaction, the only way to monitor the curing process is to measure the change in the modulus of elasticity. In dynamic mechanical analysis (DMA), an oscillatory stress is applied to a specimen and mechanical properties, such as its modulus of elasticity or mechanical damping characteristics, are measured. Tensile and shear modulus studies are used to understand the cure advancement of resin systems. In addition, conditions that could lead to phase separation of coating resins are studied by measuring the glass transition and the secondary transition and the secondary transition temperatures of materials.

Thermomechanical Analysis

Thermomechanical analysis (TMA) measures dimensional change properties of materials, e.g., expansion coefficient of materials. With this technique, a material is heated at a specified rate and relatively simple instrumentation (mechanical, optical, or electrical transducers) monitors the dimensional changes.

Table 2   Disk drive materials suitable for thermal analysis

Material	Information needed	Thermal analysis technique
All polymeric materials	Outgassing	TGA
	Thermal stability	TGA, DSC
Resin systems of particulate media	Resin/magnetic particle ratio	TGA
		DSC, DMA
	Curing (temperature, time, environment)	DMA, DSC
	Phase separation	
Lubricants	Outgassing	TGA
	Thermal stability and estimated life time	TGA, DSC
Gaskets (e.g., UV-cured material)	Effect of filler on stability and	TGA
	outgassing	TGA
	Effect of photoinitiator content, environment on outgassing and stability	
Adhesives	Cure condition in production	DSC, DMA
	Degree of cure	DSC
	Outgassing and stability of cured adhesives	TGA
Photoresists	Thermal characteristics	DSC, DMA, TGA
Metal substrate for coatings and adhesives	Thermal expansion	TMA

The linear expansion coefficient of aluminum (between 20 and 200 °C, or 68 and 390 °F) is about 25 μm/m · °C. Because drive temperatures can vary from 10 to 65 °C (50 to 150 °F), the head arm expansion should be matched with the expansion of the disk to enable the head to seek the same track at both temperature extremes. Disks are made from wrought aluminum alloy, whereas, due to cost considerations, the head arms and other parts are made from a cast alloy. The thermal expansion coefficients of both materials should be taken into consideration when the drive is designed.

In a composite head, the ferrite core is encased in a ceramic head pad material such as calcium titanate. Thermal expansion mismatch will exert stress on the ferrite core and can distort the output signal.

In thin film heads, permalloy and isolation alumina layers are deposited on a composite ceramic such as titanium carbide-alumina. Differential thermal expansion could crack the deposited layers and distort the signals generated in the permalloy. All of these considerations highlight the importance of proper selection of materials and processes for heads and disks.

In addition to measuring expansion coefficients, TMA is also used to determine the softening temperature of materials. This information is needed to establish compatibility between substrates and coatings; softening of either could lead to a permanent set or deformation.

Conclusion

High-performance drives need new and innovative materials, the properties of which are thoroughly understood. Thermal analysis techniques are important in studying many temperature-dependent properties of such materials as well as identifying materials-related potential contamination problems (Table 2). Advancing techniques and instrumentation will increase the role of thermal analysis in materials and process design as such work becomes more and more crucial to the manufacture of drives with high quality and reliability.

References

1. E.A. Turi, *Thermal Characterization of Polymeric Materials*, 2, Academic Press, San Diego, 1981.
2. W.W. Wendlandt, *Thermal Analysis*, 3rd ed., John Wiley & Sons, New York, 1986.

Appendix 1.2

Coatings for Aluminum Disk Drive Parts*

Aluminum is widely used in the computer peripherals industry because it is nonmagnetic, has excellent ductility and thus is very formable, is relatively inexpensive, and has a high strength-to-weight ratio.

When a surface finish coating is used in a disk drive, the first criterion is that it must not cause particulate or hydrocarbon contamination. Particulate microcontamination can scrape the magnetic layer or destabilize the flying head, leading to a catastrophic loss of recorded information. Hydrocarbons can affect static friction and other tribological properties in the head/disk interface.

In addition to microcontamination, many parts require electrical grounding to prevent electrical charge buildup. This makes conductivity an important criterion.

Chromate conversion coatings are often used in traditional aluminum applications. The coating process is simple and inexpensive. If the coating is scratched or damaged, the chromium seeps into that area and the coating repairs itself. The thin, conductive coating does not increase resistivity significantly.

Chromate loses many of its useful properties, including self-healing, at prolonged temperatures above 70 °C (160 °F). It also slowly dissolves in aqueous environments. Some of the critical parts in disk drives go through complex processing steps, such as deionized water cleaning, that can damage the coating.

---

* P.B. Narayan and A.S. Brar, *Mater. Eng.*, Feb, 1989, p 55. Reprinted with permission.

Some parts in high-bit-density drives have stringent contamination specifications that cannot be met by chromate conversion coatings, especially on cast aluminum surfaces.

Anodizing is another inexpensive process that provides excellent wear resistance. Plating conditions can be modified to obtain a pore-free hard coat or a conventional anodized layer for coloring.

In disk drives, parts requiring wear resistance, such as bearings, are anodized. However, even the hard coat can generate particulate microcontamination from its top layers. To solve this problem, about 10% of the anodized layer is lapped away before using the part.

Anodizing works well for simple shapes, but its uniformity is not satisfactory for complex shapes. Anodized layers are also insulators, and may lead to charge buildup problems.

Organic coatings, either electrostatic spray coated or electrostatic dip coated, work well to contain particulate contamination. However, these coatings must be tested for outgassing in disk drive environments, which can be 70 °C (160 °F) or higher.

Almost all organic coatings are electrically nonconducting, so it is necessary to mask or machine the coating off some areas to provide electrical grounding. The bare areas must then be protected with chromate conversion coating. Chromate conversion coatings improve adhesion of organic coatings and thus are often used as undercoats.

Electroless nickel is widely used on aluminum for corrosion and wear resistance, EMI/RFI shielding, and contamination control. It is nonmagnetic, amorphous, hard, and conducting. It forms a uniform thin layer even on complex shapes. It can be plated to specified thickness, and its hardness can be varied by heat treatment. However, nodular growth in electroless nickel is a significant problem.

Chemical etching, widely used to prepare aluminum for electroless nickel, weakens surface aluminum grains and intermetallic compounds, leading to particle generation. It is preferable to use mechanical means such as tumbling, bead blasting, etc., to clean the surface, followed by very light chemical etching.

Chapter 2

Analytical Instrumentation

The field of analytical instrumentation has witnessed many innovative developments in the recent past. Because the size of electronic devices currently being used is shrinking radically, the thin films used in them have thicknesses of only a few tens of nanometers, thus making the study of surfaces and interfaces very important. Also, the identification of organics on a micro- and nanoscale is becoming particularly important. Table 2.1 lists commonly used analytical procedures and instrumentation that are used in many of the case studies that appear in other chapters of this book.

Mechanical profilometers have been used for several decades to study surface topography, and their resolution capability has increasingly improved. Of late, optical profilometers have become the non-contact method of studying three-dimensional surface roughness. Scanning tunneling microscopy (STM) and atomic force microscopy (AFM) are being used in R & D laboratories routinely; however, their use in failure analysis is just emerging. Consequently, most of the case studies included in this volume did not involve the use of these ultra-modern techniques, and they are not discussed in detail.

Instead, this chapter will concentrate on the techniques used in the case studies. The failure analysis methodology and logical reasoning needed to reach a solution are the same in most of the failure case studies presented, regardless of the level of sophistication of the instrumentation used. Because of the expense involved, it is preferable to use as simple instrumentation as possible to determine the cause of failure. Table 2.2 summarizes the analytical instrumenta-

Table 1   Analytical Techniques
The following analytical instruments (with their acronyms) were routinely used in the case studies described in this volume.

AA	Atomic absorption spectroscopy
AEM	Analytical electron microscopy
AES	Auger electron spectroscopy
DMA	Dynamic mechanical analysis
DSC	Differential scanning calorimetry
EDXS	Energy dispersive X-ray spectroscopy
EELS	Electron energy loss spectroscopy
ESCA	Electron spectroscopy for chemical analysis, identical to XPS
FTIR	Fourier transform infrared spectroscopy
ICP-AES	Inductively coupled plasma atomic emission spectroscopy
LIMS	Laser-induced mass spectroscopy
LRS	Laser Raman spectroscopy
LVSEM	Low-voltage scanning electron microscopy
MS	Mass spectroscopy
RBS	Rutherford back-scattering spectroscopy
SAM	Scanning acoustic microscopy
SEM	Scanning electron microscopy (or microscope)
SIMS	Secondary ion mass spectroscopy
STEM	Scanning electron transmission spectroscopy
TEM	Transmission electron microscopy (or microscope)
TGA	Thermogravimetric analysis
TMA	Thermomechanical analysis
WDXS	Wavelength dispersive X-ray spectroscopy
XPS	X-ray photoelectron spectroscopy, identical to ESCA
XRD	X-ray diffraction
XRF	X-ray fluorescence spectroscopy

tion used to study various materials properties in many of the case studies presented in other chapters.

2.1   Study of Topography and Microstructure

2.1.1   *Optical Microscopy*

In optical microscopy, visible radiation is incident on the specimen, and the image of the surface is formed by reflected radiation. Optical microscopy is nondestructive and does not require use of a vacuum. With Nomarski and dark-field imaging attachments, organic layers a few molecular layers thick on the surface of the specimen can be imaged. The maximum magnification is on the order of 1500×, but the depth of field is very small. Optical microscopy provides very good contrast for chemically etched metallurgical specimens to study grain size and shape distribution, grain orientation, and other microstructural details such as color.

2.1.2   *Scanning Electron Microscopy*

In scanning electron microscopy (SEM), electrons accelerated through a voltage of 20 to 30 kV are allowed to strike the surface of a specimen, and the image is formed with secondary or back-scattered electrons generated at the surface. Secondary electrons provide the best resolution and topographic contrast. Back-scattered electrons provide compositional contrast based on the atomic weight differences at various locations of the surface and near-surface layers. This technique has excellent depth of field and produces images with magnification as high as 100,000×. Its biggest advantage is that it can be fitted with energy-dispersive X-ray spectroscopy (EDXS) for compositional analysis.

To perform SEM, the specimen surface must be conducting. Otherwise the incident electrons accumulate near the surface, and this negative charge buildup will repel the incident electron beam. To fulfill this requirement, in the study of insulating surfaces, a thin layer of carbon, gold, or gold-palladium alloy is deposited on the surface. Scanning electron microscopy is often a destructive technique, because of electron damage to the specimen, conducting layer deposition, and the need to fit the specimen into the SEM chamber.

Scanning electron microscopy is not well-suited to study hydrocarbon layers, which could evaporate or decompose during electron bombardment. Because good vacuum is needed to reduce electron scattering, the equipment is expensive and time consuming to use.

2.1.3   *Low-Voltage Scanning Electron Microscopy*

The advantages of low-voltage scanning electron microscopy (LVSEM) over conventional SEM include better topographic contrast, less specimen charging, and less specimen damage. The disadvantages are low source brightness, chromatic aberration, and sensitivity to stray electrostatic and magnetic fields. See Appendix 2.2 for further details regarding LVSEM.

2.1.4   *Transmission Electron Microscopy*

Transmission electron microscopy (TEM) is the only convenient technique for studying the internal microstructure of thin films and bulk materials at magnifications as high as 500,000×. In the development of thin films, microstructural studies with TEM provide invaluable information about the nucleation and growth processes. Both TEM and SEM are complementary techniques that are essential for good microstructural understanding of all materials, particularly thin films.

In TEM, electrons accelerated to 200 kV (and sometimes as high as a million volts) pass through the bulk of the specimen, undergoing various elastic and inelastic, coherent and incoherent scattering processes to form an image. Electron diffraction and scattering by the microstructural features provide contrast in the transmitted electron image. Transmission electron microscopy does not experience the charging problems of SEM and provides the best resolution possible.

In the TEM mode, selected area electron diffraction patterns (SAEDP) of the specimen area, as small as 100 nm (1000 Å or 3.94 μin.), can be obtained to determine its crystalline structure. The patterns can be indexed to obtain information about the lattice spacing and orientation. It is an invaluable technique for studying nucleation or precipitation of new crystalline phases. Appendices 2.3 and 2.4 describe the advantages of TEM application in studying particulate oxide media.

Because electrons must travel through the specimen thickness, to produce adequate image brightness the specimen must be thinner than 100 nm (3.94 μin.). Preparing such a thin section of the specimen is the most difficult task in TEM application. Also, it is difficult to obtain information from a specified location. Consequently, TEM is used to study microstructure in random areas.

The SEM and TEM techniques can be combined to obtain scanning transmission electron microscopy (STEM). As in SEM techniques, the specimen is rastered with an electron beam. However, the sample is so thin that the transmitted electrons form an image. The advantage of STEM over TEM is that the focused electron beam in the STEM mode can be used to form the diffraction pattern (called the micro-microdiffraction mode) of very fine precipitates of the size of 10 nm (0.394 μin.).

2.1.4.1 Specimen Preparation Methods

Chemical Thinning. The material under study is chemically dissolved in an acidic or alkaline solution to obtain a thin specimen. This is not a well-controlled process, although it is the easiest to undertake. Different phases in the specimen will have different dissolution rates, leading to nonuniform specimen thickness.

Electrolytic jet polishing can be used only for specimens of conducting materials (e.g., metals). Voltage is applied through an electrolytic jet to the specimen, thereby electrochemically dissolving the surface layers. As soon as a tiny hole develops in the specimen, a laser beam activates a diode, thereby terminating the voltage application. The area around the hole is thin and transmits electrons. This method of sample preparation is fast, inexpensive, and well controlled. It is ideally suited for bulk alloys.

Ultramicrotomy. The specimen is embedded in an organic resin and sectioned with a diamond knife. The thickness of the section can be as little as 10 nm (0.394 µin.). The major advantage of ultramicrotomy is that it provides a large area of uniform thickness relatively quickly. See Appendix 2.5 for the use of ultramicrotomy in the analysis of magnetic recording media.

This technique is useful for organic coatings, membranes, and relatively soft metals and materials. Ceramic materials cause excessive wear on diamond knives. Because it is a mechanical sectioning process, use of ultramicrotomy can cause stress in the specimen and can generate dislocations and other crystalline defects, especially in metals.

Ion Milling. In this method, argon ions abrade the surface of the specimen. As soon as a hole develops, a laser beam terminates any further milling, similar to the electrolytic jet polishing technique. It is the only technique available for making thin sections of ceramics. Ion milling is versatile and can be used on all materials.

During processing, the specimen becomes heated, often to a few hundred degrees Celsius. If the specimen is likely to undergo thermal damage, it can be cooled with liquid nitrogen. This technique is time consuming, often requiring several days to obtain a thin section.

2.1.5 *Scanning Acoustic Microscopy*

With scanning acoustic microscopy (SAM), ultrasonic radiation is incident on the specimen, and the image is formed by the reflected ultrasonic waves. It is a nondestructive method of studying subsurface defects in ceramics, such as delamination, cracking, and voids.

2.2 Study of Crystallography

2.2.1 *X-Ray Diffraction*

When monochromatic X-rays are incident on a specimen, they undergo diffraction according to the Bragg rule:

$$\lambda = 2d_{hkl} \sin\theta$$

for first-order diffraction.

Here λ is the wavelength of the X-ray radiation, d_{hkl} is interplanar spacing for the *hkl* plane of the specimen, and θ is the angle the incident beam makes with the crystallographic plane. Because λ is known and remains constant for a particular monochromatic radiation, by measuring θ, the interplanar spacings, d_{hkl} can be calculated. From d_{hkl} values, the lattice parameter, a, can also be calculated. For example, for cubic structures:

$$d_{hkl} = a/\sqrt{(h^2 + k^2 + l^2)}$$

Because stress changes the lattice parameter, by measuring the lattice parameter, the retained stress in the specimen can be estimated. Because X-rays are hard to focus, the lateral resolution is on the order of several millimeters. The depth of resolution is several micrometers. Several modifications (e.g., grazing incident angle, rocking curve, etc.) exist for thin-film studies.

2.2.2 *Selected Area Electron Diffraction Patterns*

Selected area electron diffraction patterns (SAEDP), in the TEM mode, are also used to determine crystalline parameters. The principle used is the same as for X-ray diffraction (XRD). Crystal structure determination with SAEDP is much more cumbersome than with XRD, but its advantage lies in being able to identify the crystallinity of small areas (a tenth of a micrometer) in the TEM image.

2.3 Study of Chemical Composition

Table 2.3 shows the incoming and outgoing radiation for various techniques used for measuring chemical composition. The wet chemical techniques of atomic absorption spectroscopy (AA), inductively coupled plasma atomic emission spectroscopy (ICP-AES), and mass spectroscopy (MS) involve dissolving the material in a suitable solvent and precisely measuring its chemical composition based on given standards. These are bulk techniques and are not well-suited for local elemental identification.

Energy-dispersive X-ray spectroscopy and wavelength dispersive X-ray spectroscopy (WDXS) are often used in conjunction with SEM. They have good lateral resolution on the surface. However, the X-ray generation process occurs in a spherical area inside of a specimen with a diameter of several micrometers. The diameter of the spherical area becomes smaller as the atomic number of the material increases. In EDXS, the energy of X-rays (generated by the electron irradiation of the specimen) is measured. This technique is fast, and its detection limits and accuracy are within ±2 to 5%. With WDXS, the wavelength of the X-rays is measured. It is a slower technique than EDXS, but its accuracy and detection limits are better.

Auger electron spectroscopy (AES) and electron spectroscopy for chemical analysis (ESCA) are surface-sensitive techniques in which the depth of resolution is a few nanometers. Used in combination with argon ion milling, these techniques facilitate depth profiling, and three-dimensional compositional mapping is possible.

In AES, the incoming electrons cause Auger transitions in the elements present in the specimen. The Auger energy levels are characteristic of each element, and the presence of an element can be determined by measuring the electron energies. Because electrons can be focused easily, AES has the best lateral resolution, and the spot size can be as small as 10 nm (0.394 μin.). Its ability to provide information about the chemical bonding state of an element is limited, and its detection limits and accuracy are approximately ±2%. Its resolution is best suited for lower atomic number elements. When studying nonconducting specimens, charging problems have been experienced.

In ESCA/X-ray photoelectron spectroscopy (XPS), a beam of monochromatic X-rays incident on the specimen knocks out the orbital electrons. By measuring the kinetic energy of the electrons, their binding energy (which is a characteristic of an element) can be estimated. The spot size is several hundred micrometers. Electron spectroscopy for chemical analysis also provides information on the chemical bonding state. This technique is more accurate and experiences fewer charging problems than AES.

Fourier transform infrared spectroscopy (FTIR) is ideally suited for determining the organic bonding state of thin hydrocarbon layers nondestructively. Appendix 2.1 discusses the various thermal analysis techniques of thermogravimetric analysis, thermomechanical analysis, differential scanning calorimetry, and dynamic mechanical analysis.

Appendix 2.1

Low-Voltage SEM (LVSEM) in the Nondestructive Analysis of Computer Peripheral Products*

As computer memory technology advances, there is increasing demand for higher information storage densities in magnetic recording disk drives, while at the same time, their physical size needs to be decreased. To accommodate higher bit and track densities (the product of which determines the storage density of the disk), the magnetic medium on the disk is often thinner than 0.5 µm, and the read/write head with a submicrometer gaplength flies above the disk at heights lower than 0.3 µm.

Because media parameters such as magnetic layer thickness, head flying height, and head gaplengths have submicrometer dimensions, microcontamination alone can adversely affect the highly fragile and intricate head/disk dynamics, and thus, the identification and control of contamination are crucial to ensure the reliability of disk drives. Consequently, scanning electron microscopy (SEM) is being used increasingly in the computer peripheral industry to study microcontamination, disks, heads, and other components.

Particulate iron oxide is the primary choice of magnetic material for information storage at present. Needlelike gamma iron oxide, about 0.7 µm long and with an aspect ratio of about 7:1, is milled along with a suitable surfactant to obtain the optimum dispersion and is then mixed with an organic binder resin to form a magnetic paint. A few percent of hard alumina also is added to the magnetic paint to improve the wear resistance of the disk during its interaction

---

* A.S. Brar and P.B. Narayan, *Microelectronic Manufacturing and Testing*, July, 1987. Reprinted with permission.

with the recording head. The magnetic paint is spin coated onto a well-polished and diamond-turned aluminum substrate (in the same basic manner that it is applied to a wafer), and the resin is thermally cured.

In the current "Winchester" drive technology, the head rests on the disk, and when the drive is turned on, the head lifts up due to the aerodynamic thrust provided by the disk rotating at 3600 rpm. Similarly, when the drive is turned off, the head lands on the disk. To reduce wear at the head/disk interface during the head take-off and landing, a very thin film (about 40 Å, or 4 nm thick) of a fluorocarbon-type lubricating fluid is applied on the disk, either by dip coating or by a spray and buff technique.

As the disk is rotating at 3600 rpm during the drive operation, the lube tends to spin off. If the lubrication coating becomes too thin, there is unacceptable wear and friction at the head/disk interface. If the lube is too thick, stiction (static friction) will increase, thus causing the head to stick to the disk. However, if there is controlled porosity in the magnetic coating, the pores act as lube reservoirs and supply lube to the head/disk interface at the desired rate. Thus, it is very important to monitor the surface porosity of the magnetic coating and the distribution of the lube on it, using an SEM.

Two types of heads—thin film and ferrite (either NiZn or MnZn)—are widely used at the present time. The thin-film heads (usually made of Permalloy) are sputtered on to a ceramic slider material (e.g., Alsimag), whereas the ferrite cores are encased in a suitable ceramic (e.g., Fotoceram, calcium titanate, etc.) using a bonding glass of low melting point. The submicrometer-size gaplength usually is obtained by a sputtered layer of aluminum oxide. The dimensions of the gap length, the smoothness of the slider, and the presence of microcontamination on the heads are some of the parameters that should be monitored constantly with an SEM during head manufacturing.

Role of LVSEM

In SEM, a finely focused beam of electrons that are accelerated through a chosen potential gradient scans the specimen surface in a raster pattern. In an ideal case, the emitted secondary electrons form the usual SEM image in the scanning electron imaging (SEI) mode. In reality, in addition to the secondary electrons produced at the electron beam/specimen contact area, back-scattered electrons, secondary electrons that are emitted when back-scattered electrons come out of the specimen, and the secondary electrons produced when either incident or back-scattered electrons strike pole pieces, apertures, etc., also contribute to the SEM image (Fig. 1).

All of the emitted electrons, other than the secondary electrons produced at the beam/specimen contact area, tend to decrease the lateral and depth resolutions of the image. Secondary electrons have low energy (0 to 50 eV) and can travel only a few decades of angstroms in solids. Thus, the SEM image produced only with the secondary electrons generated at the beam/specimen contact area will have the best resolution and topographic contrast.

Typically, at 15 keV, about 20% of the incident electrons become back-scattered, and the lateral and depth resolutions are about 1 to 1.5 μm. There is about 20% efficiency in producing the secondary electrons at the beam/specimen contact area.

The total electron emission, including the secondary and the back-scattered, is significantly lower than unity at the accelerating voltages of 20 to 30 keV and rapidly increases as the accelerating voltage decreases. After crossing unity, the emission reaches a maximum at around 1 to 1.5 keV and then starts decreasing with a further decrease in voltage (Fig. 2). Thus, at higher voltages, there is a negative charge buildup, which severely distorts the incident beam and the image signal.

A conventional SEM is generally operated at 20 to 30 keV to obtain the best signal-to-noise ratio and a clear image. Ceramic slider materials, organic resin coatings (e.g., in the particulate oxide disks), and ferrites are relatively nonconducting and hence are especially prone to charging problems when studied with conventional SEM. To alleviate the charging problem, the insulting specimens are often coated with a very thin film of gold, palladium, or carbon. However, this procedure is destructive and is thus unacceptable in device processing. In failure analysis, the overcoat also might interfere with the evidence and thus should be avoided. Overcoating the specimen for SEM examination can be avoided at low voltages (1 to 5 keV) because there is a positive charge buildup on the specimen surface at these low voltages, thereby minimizing charging problems.

Exposing organic films such as the lube to high-energy (20 to 30 keV) electrons can lead to polymerization and ionization damage. Because the low-voltage electrons have much less energy, they cause considerably less specimen damage. Also low voltages have a higher cross section for the production of secondary electrons, and hence, the SEM image becomes more surface sensitive, thereby improving the topographic contrast. For example, the ranges of electrons for silicon are 0.032 μm at 1.0 keV and 9.3 μm at 30.0 keV. Thus, low-voltage scanning electron microscopy (LVSEM) can play a vital role in studying computer peripheral products, especially the heads and the disks.

Advantages and Disadvantages of LVSEM

To summarize, the advantages of LVSEM are less specimen charging, less specimen damage, and better topographic contrast. Its main disadvantages are low source brightness (could be improved by using LaB_6 and field emission electron guns), chromatic aberration (could be improved by decreasing the focal length of the lenses, and by using LaB_6 or FE guns), and sensitivity to stray electrostatic and magnetic fields (could be improved by better shielding and better design).

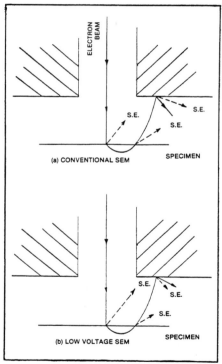

Fig. 1 Secondary electron emission mechanism in conventional (a) and low-voltage (b) SEM. Length of the dashed arrows approximately represents the intensity of the secondary electrons.

Study of Magnetic Media by LVSEM

The magnetic and lubrication layers on the particulate oxide disk are nonconducting and have a relatively low atomic number and thus provide very little contrast in conventional SEM. However, with LVSEM one can clearly see surface porosity with significantly better topographic contrast (Fig. 3). The bright areas in and around the pores indicate the presence of the lubrication fluid. It appears that lube is not present as a uniform thin film, but it has much higher thickness at the pores. The pores in the coating act as lube reservoirs and supply lube at the head/disk interface. This type of study is

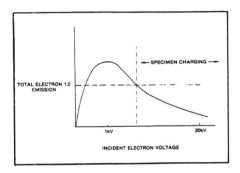

Fig. 2 Total electron emission (secondary and back-scattered) as a function of incident electron voltage.

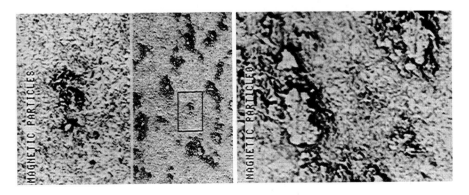

Fig. 3 Surface porosity on a particulate oxide disk. Bright areas near the pores indicate the presence of lubrication fluid. Pores act as lube reservoirs.

Fig. 4 Aluminum substrate showing the presence of intermetallic compounds that tend to produce missing bits and errors during recording.

helpful in optimizing the surface porosity and distribution of lube on the surface.

Study of Aluminum Substrates

Rigid disk substrates are made with an aluminum-magnesium alloy, which generally has intermetallic compounds of the type AlFeSi. The intermetallics do not get coated by the magnetic paint very well and consequently show up as missing bits and errors during recording. If the intermetallics are very finely dispersed on the alloy surface, they may not pose as significant a problem as when they are agglomerated. When imaged in a SEM, they tend to get charged at high voltages, thus losing all resolution, but they can be clearly seen with LVSEM (Fig. 4). High-purity alloys tend to have fewer intermetallics and thus produce fewer errors during recording.

Study of Ferrite Composite Heads

Figure 5(a) shows a ferrite core in a composite head, and the gapwidth can be accurately measured from it. There is very little charging at these low voltages, and thus, all the features such as ferrite grain size can be clearly observed. The brighter grain probably has a different orientation and thus was polished differently. Porosity in the core material can be clearly observed. From Fig. 5(b) (a high-magnification version of Fig. 5a), the gaplength can be measured accurately.

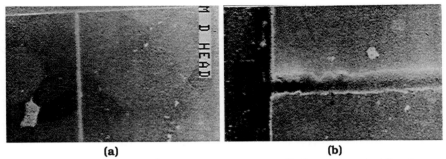

(a)   (b)

Fig. 5 Manganese-zinc ferrite core in a composite head showing the gap and the grain structure of the ferrite (a). The gaplength can be measured accurately from (b), which is a high-magnification view of (a).

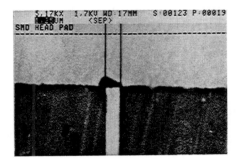

Fig. 6 Ferrite core of a failed composite head. The undesirable ferrite (or contamination) particle across the gap might have caused the failure of the head.

Failure Analysis

Figure 6 shows a failed ferrite composite head with an undesirable particle short-circuiting the gap, which would adversely affect the head performance. The deep polishing lines on the ferrite are clearly visible. Such rough polishing would create a deformed layer on the ferrite core surface, adversely affecting its magnetic properties.

Study of Organic Contamination

Organic contamination due to human handling, atmospheric hydrocarbons, residues left over from various processing steps, etc., are a major concern in the manufacture of computer peripherals and thus should be closely studied. If the organic contamination is exposed to high voltages, it becomes decomposed thus forming a deposited layer of carbon. However, it too can be studied and photographed at low voltages relatively easily.

Conclusion

Until now, the application of LVSEM has been confirmed largely to the study of photoresists, nonconducting silicon wafers, and other materials in the semiconductor industry. The present work indicates that it can be used effectively in the computer peripherals industry as well, to study the particulate iron oxide and sputtered metallic thin-film disks, aluminum substrates and their intermetallics, and recording head surfaces. Use of LVSEM contributed significantly in understanding the distribution of the lubrication fluid on the oxide disk.

Although the technique needs many improvements, it has been found to be very helpful in the failure analysis of magnetic media and heads in disk drives. It should be realized that LVSEM operations require a lot of experience and dedicated effort. With the anticipated decreases in media parameters, LVSEM will continue to have an effective role in the research and development of computer peripherals.

References

- J. Pawley, "Low Voltage Scanning Electron Microscopy," *J. Microscopy,* Vol 136, 1984, p 45-68
- M.T. Postek and D.C. Joy, "Microelectronics Dimensional Metrology in the Scanning Electron Microscopy," *Solid State Technology,* Nov, 1986, p 145-150
- D.F. Blake, "Low Voltage Scanning Electron Microscopy," *Test & Measurement,* June, 1986, p 62-75

Appendix 2.2

TEM Divulges The Innermost Secrets
of Your Recording Media*

As computer memory technology advances, the bit and track densities of magnetic recording media are increasing dramatically. The most limiting constraints to high bit density longitudinal recording are the recording demagnetization and self- and adjacent-bit demagnetization processes.

To minimize the problems associated with these demagnetization processes, the medium should be very thin (6000 Å) and have high coercivity and low magnetization. Thin, low-magnetized media lead to extremely small signals that, in turn, require a read/write head flying height of ~3000 Å. This extreme scaling of recording media parameters tends to magnify the impact of imperfections and defects. Thus, it is very important that the media constituents be studied in much closer detail than before.

Memory storage particles that are used in particulate recording media include iron oxides (gamma Fe_2O_3 and magnetite), cobalt-modified (cobalt-substituted and cobalt-absorbed) iron oxides, barium ferrite, chromium dioxide, and a variety of metal particles. Gamma iron oxide is the most commonly used of all the particles and is manufactured from ferrous sulfate through a series of oxidation, reduction, and dehydration steps.

The iron oxide used in longitudinal recording media generally consists of acicular (with a length-to-width ratio between 3:1 and 10:1) single-domain particles (1000 to 7000 Å long). Depending on

---

* P.B. Narayan and A.S. Brar, *Research & Development*, May, 1985, p 121-126. Reprinted with permission.

the magnetic requirements of a particular head/disk combination, an iron oxide of suitable coercivity and physical characteristics is chosen.

The other submicrometer media constituents are the load-bearing ceramic particles. They provide wear resistance for the relatively soft media during interaction with the read/write head. Alumina (Al_2O_3) is the most commonly used ceramic particle. The alumina particles generally are rounded and have dimensions near the media thickness.

To remove agglomeration and obtain good particle dispersion, both iron oxide and alumina particles are milled for a predetermined length of time, mixed with a binder resin, and coated on the substrate—aluminum alloy for rigid disks and plastic for flexible disks. Rigid disks are passed through a magnetic field that aligns (orients) the iron oxide particles in the direction of the magnetic field. The resin is then cured, and the disk is polished.

Electron microscopy is the most effective direct imaging technique for studying the submicrometer-size constituents of a magnetic medium (Fig. 1). Scanning electron microscopy (SEM) is best used for acquiring surface topographic information. Transmission electron microscopy (TEM) is most effective at revealing the geometric aspects of particles with good clarity and resolution at high magnifications (300,000× and above). The two microscopic techniques often are complementary, but it is the information provided by TEM that has been most useful in studying various aspects of magnetic recording.

In TEM, the specimen thickness through which the transmitted beam can pass without a significant loss in energy and brightness increases as the accelerating voltage (kinetic energy) of the electrons increases. For 100-kV electrons, the maximum specimen thickness is 1000 Å. As a result, TEM specimen preparation techniques can be complex and challenging.

Areas in which TEM has been of immense help in producing high-quality recording media include determining iron oxide and alumina quality, particle dispersion, particle distribution, iron oxide orientation, and surface porosity.

Iron Oxide and Alumina

Before an iron oxide source can be chosen, samples from various vendors are studied with TEM. Physical characteristics that are

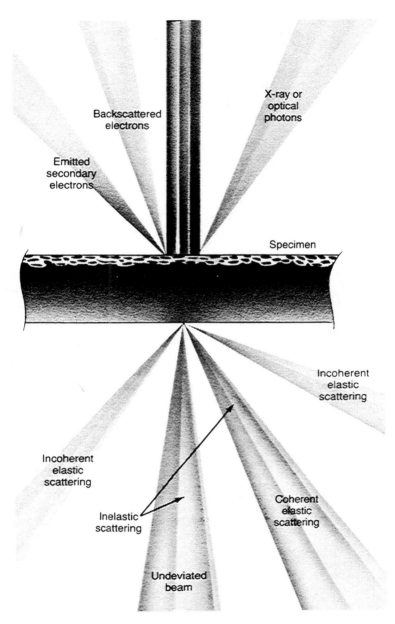

Fig. 1 TEM images are generated by electrons that are transmitted through a specimen from an incident beam. These images have proven very helpful in assessing recording media quality.

investigated include particle size, aspect ratio (ratio of particle length to width/diameter), size and shape uniformity, acicularity, presence of holes and dendrites, and ease of dispersion.

All of these characteristics have a strong effect on magnetic properties, because coercivity generally decreases with an increase in particle size. The orientation ratio, however, is highest when small particles with small aspect ratios are used. A high orientation ratio generally leads to a high degree of hysteresis loop squareness and a narrow switching field.

It is important to have a consistent distribution of both particle size and aspect ratio (Fig. 2). An even particle dispersion leads to comparatively better and more reliable magnetic properties, whereas the presence of dendrites hinders oxide particle orientation (Fig. 3). By studying the iron oxide powder, it is possible to reliably predict the media quality that will result from a particular oxide.

Just as with iron oxide, the various physical characteristics of alumina, i.e., particle size, shape, size and shape distributions, and ease of dispersion, can be studied with TEM. The load-bearing alumina particles prevent inadvertent contact between the head and the medium and extend the life of the disk.

Alumina particle roundness is inversely proportional to the number of errors (extra bits) that occur. Particles with dimensions equal to the medium thickness generally provide good wear resistance for

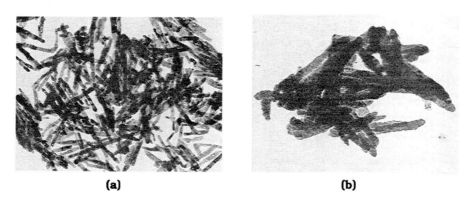

(a)   **(b)**

Fig. 2 Iron oxide is milled to break up agglomerates and provide well-dispersed, easily oriented particles (a). Insufficient milling time results in poorly dispersed particles (b).

the medium, whereas particles that are too small do not contribute to wear resistance but do increase error occurrences.

Particles that are too large may project out of the medium and interfere with the flight of the read/write head because the head flies within 3000 Å of the surface. Thus, it is important to be able to disperse the alumina evenly to achieve uniform protection over the entire disk.

Dispersion and Distribution

Particle milling time must be carefully controlled because too short a time leads to agglomerated iron oxide and alumina on the disk, whereas too long a time leads to particle damage and contamination from the milling apparatus. A TEM will provide a clear visual indication of the degree of dispersion and the degree of particle damage that has occurred during milling.

Iron oxide particles are extracted from the medium on the finished disk and their size and other physical parameters studied using TEM to ensure that no particle damage occurs during disk coating and polishing. Alumina also is checked for uniform dispersion.

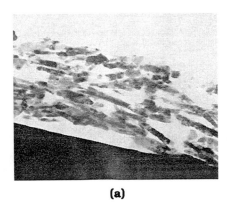 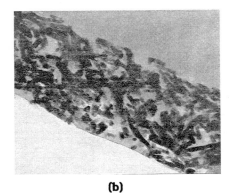

(a)   (b)

Fig. 3 Soon after a rigid disk is coated, it is passed through a magnetic field to orient the iron oxide particles. TEM images of disk cross sections show good orientation (a) and poor orientation (b).

(a)   (b)

Fig. 4 Pores in recording media surfaces act as reservoirs for lubrication fluid. Surface porosity can be investigated using negative replication and TEM. Pictures show large pores (a) and small pores (b).

Orientation and Porosity

Transmission electron microscopy cross sections of a disk in two perpendicular directions reveal the degree of iron oxide orientation. Strong orientation gives high resolution and a high signal-to-noise ratio during data storage and retrieval. Uniform magnetic properties in the medium can be checked by studying the oxide orientation at the outer, middle, and inner diameters of the disk.

These TEM orientation studies may be used as a complement to studying disk magnetic properties using a vibrating sample magnetometer. Transmission electron microscopy cross section studies also reveal the alumina location in the media.

Transmission electron microscopy, in combination with negative replication techniques, can be used to study surface porosity after polishing (Fig. 4). Optimum porosity is required because the pores act as a reservoir for lubrication fluid and ensure its supply at the head/medium interface during the life of the memory storage drive.

Transmission electron microscopy is becoming increasingly important in the study of various aspects of magnetic media to optimize processing parameters during the manufacture of rigid oxide media. This leads to the production of higher quality media for use with current and future generations of high-density computer disk memory devices.

Appendix 2.3

Using SEM and TEM to Analyze Rigid Disks*

Because of their durability, rigid disks with particulate magnetic media are favored for high bit density computer memory applications. These media consist of aluminum disks coated with a mixture of gamma iron oxide and aluminum oxide (alumina) particles suspended in a resin binder. The iron oxide, being a permanently magnetic material, is the data storage material. The alumina, because of its hardness, adds durability to the disk. High signal amplitude and good signal resolution determine a good disk. Stronger orientation, better dispersion, and tighter size distribution of the iron oxide particles improve disk quality. Missing or extra bits of encoded data, on the other hand, are potential error sources that well-dispersed alumina can help minimize.

One of the main threats to head disk longevity is head crash, in which the read/write head (usually made of a hard ceramic material with ferrite) inadvertently touches the softer disk media. Kawakubo et al. (Ref 1) propose a model for head crash in which the head picks up disk/binder wear particles, thus coming in contact with the disk. They suggest that alumina in the media can clean the head by knocking off debris and so prevent, or at least delay, head crashes.

Disk manufacturers mix needle-shaped iron oxide particles 0.1 to 0.7 μm long with surfactant and binder resin, and then ball or sand mill the mixture to achieve good particle dispersion. Alumina particles (about a micron in size) undergo similar treatment. After combining the iron oxide and alumina mixtures, the manufacturer coats

---

* P.B. Narayan and A.S. Brar, *Test & Measurement World*, Vol 6 (No. 2), Feb, 1986, p 54-60. Reprinted with permission.

a polished aluminum alloy disk with the resulting magnetic paint
and passes it through a strong magnetic field to orient the oxide
particles. After the resin cures, a final polishing step brings the
coated disk to its required thickness and surface finish.

Etching Disks for SEM and TEM Analyses

Combined with electron microscopy, plasma etching allows studying the orientation and physical characteristics of magnetic oxide
particles. Chemical etching allows studying alumina particle distribution on the magnetic oxide disk.

Plasma etching entails placing a sample in a chamber filled with
plasma so that the rapidly moving gas ions strike the sample,
sputtering off surface atoms. Plasma etches away the sample surface
in a very predictable manner. Chemical etching etches the sample
surface by contact with acids or other corrosive chemicals. Like
plasma etching, chemical etching is highly predictable. Oxygen-
plasma etching combines these techniques. Oxygen ions are extremely reactive chemically, quickly burning away resin molecules.
Because both disk-coating components, iron oxide and alumina, are
relatively hard oxides, oxygen-plasma etching can effectively remove

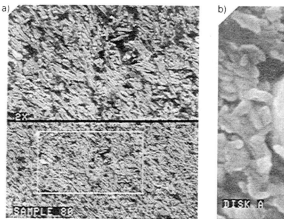 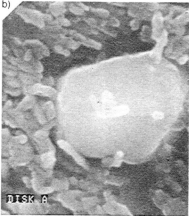

Fig. 1 SEM micrographs showing (a) magnetic iron oxide orientation on a
plasma-etched rigid memory disk. (b) Location of alumina at the head/disk
interface.

the resin binder on the disk with little or no disturbance of the iron oxide and alumina particles.

For high bit density applications, the magnetic coating on the disks should be as thin as possible (Ref 2). However, extreme scaling—keeping disks as thin as 0.6 µm and maintaining flying heights as low as 0.3 µm—magnifies the harm media defects cause. Using the high resolution of TEM and the surface topographic information supplied by SEM allows studying thin rigid disk coatings in extreme detail. Sputtering off organic resin binder from the disk by oxygen-plasma etching exposes the magnetic medium for SEM study of iron oxide particle orientation (Fig. 1a) and alumina particle location at the head/disk interface (Fig. 1b). Because of the resolution limitations of SEM, TEM is better suited for studying particle size distribution. Poor size distribution of magnetic particles can affect coercivity (ability of the coating to resist demagnetization), print through, and other properties of the disk.

Preparing a TEM specimen involves plasma-etching the media, scraping the coating from the aluminum substrate, dispersing it in silicone oil, and spreading the resulting mixture over a collodion cast

0.5 micrometer

Fig. 2 TEM micrograph of magnetic iron oxide obtained by plasma etching the disk, then scraping the coating and dispersing it in oil.

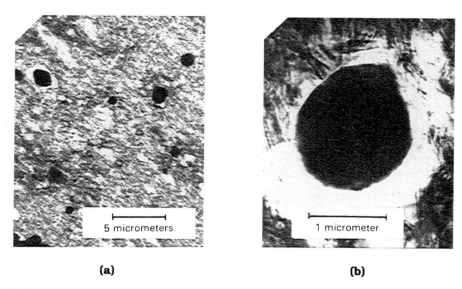

5 micrometers	1 micrometer
(a)	**(b)**

Fig. 3 TEM micrograph (a) and detail (b) of magnetic media with large, uniform and well-dispersed alumina particles.

on water. A 300-mesh copper grid supports the cast and the iron oxide particles. Figure 2 is a TEM micrograph of iron oxide from a disk showing particle size distribution.

Comparing the size distribution of the plasma-etched oxide particles with the distribution prior to milling and dispersion shows whether milling, coating, and polishing cause particle breakage. Iron oxide and alumina milling are the primary sources of ceramic debris in the media. Debris also comes from ball milling constituents—zirconia balls and alumina container linings—as they wear. This debris introduces noise into the read/write process, degrading disk performance.

Too mild a process leads to iron oxide and alumina agglomeration, which in turn degrades disk performance. Too harsh a milling process, on the other hand, can cause particle breakage and produce more debris. The particle size distribution should remain unchanged throughout the manufacturing process to ensure that disks meet specification. Studying the ceramic particle size distribution in the media by TEM and SEM analysis of plasma-etched and chemically etched disks allows altering the milling process to generate the least amount of debris.

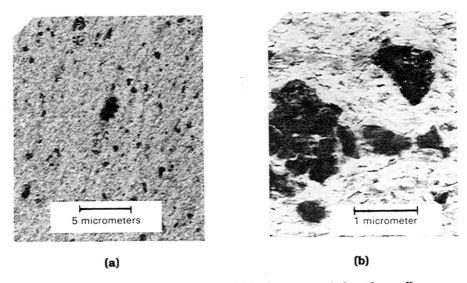

(a)   (b)

Fig. 4 TEM micrograph (a) and detail (b) of magnetic disk with smaller, less uniform and more poorly dispersed alumina than that shown in Fig. 3.

Chemical Etching and TEM Analysis

Immersion in dilute nitric acid (HNO_3) separates the coating from the aluminum substrate, but the coating is still too thick for TEM analysis. Chemically etching the coating in dilute hydrofluoric acid (HF) for a few minutes thins it enough to enable TEM observation. The etching rate depends on the properties of the binder resin system, so optimum HF concentration and etching time must be determined by trial and error.

Figures 3 and 4 compare the alumina distributions on two different chemically etched disks. The disk in Fig. 3 has larger, more uniform, and better dispersed alumina than the one in Fig. 4.

Ensuring that a rigid disk meets electrical performance standards requires examining its physical characteristics, such as particle size distribution, dispersion, and iron oxide particle orientation. Size distribution and alumina dispersion studies can help minimize disk errors and wear. Plasma and chemical etching in combination with transmission and scanning electron microscopy are complementary techniques that are indispensable for analysis of iron oxide and alumina on disks.

References

1. Y. Kawakubo, H. Ishihara, Y. Seo, and Y. Hirano, *IEEE Trans. Mag.,* Vol MAG-20, 1984, p 933.
2. D.E. Speliotis, *IEEE Trans. Mag.,* Vol MAG-20, 1984, p 669.

Appendix 2.4

Ultramicrotomy in the Analysis of Magnetic Recording Media*

Particulate magnetic recording media is widely used to store information in computer memory applications. The media consists of submicron-size gamma iron oxide and alumina particles held together by a binder resin onto a substrate (mylar for floppy disks and aluminum for rigid disks). The electrical properties of media depend on the physical characteristics (size and shape), dispersion, and orientation of iron oxide. Well-dispersed and optimum-sized alumina improves wear resistance of the relatively soft media. Transmission electron microscopy (TEM), with its excellent spatial resolution, is very useful in analyzing media (Ref 1). Ultramicrotomy can be conveniently used to prepare cross sections of the media for TEM study. The major advantage of ultramicrotomy is that it provides a large amount of uniformly thin area relatively quickly.

In a rigid disk, the aluminum substrate is much harder than the media and thus more difficult to cut, even with a diamond knife. If physical characteristics of iron oxide and alumina need to be studied, the media can be separated from the aluminum substrate by immersion of the disk in dilute nitric acid (Ref 1). The separated media can then be fixed in an embedding resin and cross sectioned with a diamond knife in an ultramicrotome. Media separated from aluminum tends to curl, making it difficult to infer oxide orientation

---

* P.B. Narayan, in *Proceedings of the 43rd Annual Meeting of the Electron Microscopy Society of America*, G.W. Bailey, Ed., San Francisco Press, 1985. Reprinted with permission.

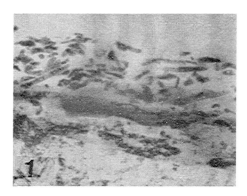

Fig. 1 TEM micrograph of circumferential cross sections of a rigid disk.
Under the iron oxide coating is the aluminum substrate.

Fig. 2 TEM micrograph of radial cross section of a rigid disk.

from the cross section. The aluminum substrate keeps the latter
unchanged during sectioning. In the present investigation, it was
found that a few microns of the aluminum substrate would keep the
oxide orientation intact and at the same time would not cause
excessive wear on the diamond knife.

Figure 1 shows the circumferential cross section of a rigid disk,
and Fig. 2 shows its radial cross section, radial being perpendicular
to circumferential. Figures 1 and 2 illustrate that iron oxide is
circumferentially orientated. The oxide particles have a diameter of
0.06 μm and a length of 0.6 μm, giving an aspect ratio (ratio of length

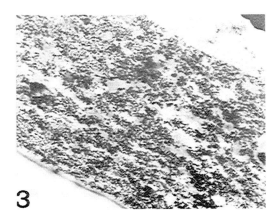

Fig. 3 TEM cross section of a floppy disk. The small rice-like particles are iron oxide and the bigger particles are alumina.

to diameter) of about 10. The media is about 0.6 μm thick. The oxide particles are well dispersed. Good dispersion and orientation of iron oxide lead to better signal resolution and signal-to-noise ratio.

Although the floppy substrate is softer than aluminum (and hence causes less wear on the diamond knife), adhesion between the media and substrate is not good, which often leads to separation of media from substrate during sectioning. Figure 3 is a TEM micrograph of the cross section of a floppy disk. The iron oxide particles are about 0.1 to 0.2 μm long and do not show any orientation. The media is about 6 μm thick, and the size of alumina is about 0.5 μm. Figure 3 illustrates that alumina is well dispersed.

Ultramicrotome cross sections of media provide much information about the size and shape of iron oxide and alumina, dispersion and agglomeration of oxide and alumina, and orientation of oxide, all of which have critical influence on the properties of the media.

Reference

1. P.B. Narayan and A.S. Brar, "Analysis of Magnetic Oxide Media in Computer Memory Disks," *Proc. 4th Ann. Test and Measurement World Expo,* Vol 1, San Jose, May, 1985, p 191-201.

Chapter 3

Protective Coatings

Corrosion is a primary mechanism by which particles are generated. All metals (except gold) can be oxidized, because metallic oxides have lower Gibbs free energy than the parent metals, making oxidation a spontaneous and irreversible reaction. For metals such as aluminum and chromium, the oxide layer is tenacious, adherent, and pore-free. The protective oxide scale, in general, exhibits low thermal expansion mismatch with the parent metal and small volume change during oxide formation.

Once the surface is covered with a few molecular layers of defect-free oxide, the oxidation reaction is drastically slowed because any further oxidation requires solid-state diffusion of metal ions through the oxide layer to the oxide-air interface. The equation below shows that diffusion exhibits an Arrhenius dependence on temperature. Consequently, the diffusion rate is very small at ambient temperatures:

$$D = D(0) \exp(-Q/RT)$$

where D is the diffusion rate, T is the absolute temperature, R is the Universal Gas Constant, Q is the activation energy, and $D(0)$ is a constant.

Although atmospheric oxygen is the primary oxidizing agent, the presence of moisture facilitates oxidation, which is always enhanced in high-humidity environments. Species such as chlorine, when present even in trace amounts, can reduce oxide in isolated locations, remove the passivating oxide layer, and expose the metal to facilitate further oxidation, thus leading to pitting.

Corrosion begins on a surface, and a protective coating can be applied to the surface without affecting substrate properties such as strength and magnetizing field strength. The salient features of protective coatings commonly used in disk drives are discussed below. Each coating has its own advantages and disadvantages. It is necessary to select the optimum coating for each part based on its function, environment, and interaction with other components.

Wrought and machined surfaces provide a uniform and homogeneous surface for coatings. Cost considerations favor replacing wrought parts with sand cast and die cast alloys. Several elements are added to the alloys to improve castability. For example, addition of 7% Si to aluminum alloys increases the fluidity of the molten metal. These added elements form intermetallic compounds that make the metal surface inhomogeneous. Mold contamination (e.g., sand particles from the sand mold), mold release agents, and die contamination, when present on the surface, could introduce defects in the protective coatings.

Some of the important considerations in selecting an overcoat are (1) corrosion resistance, (2) coating thickness when the part has tight dimensional tolerance, (3) uniformity of coating thickness when the part has complex shape, (4) electrical conductivity to facilitate charge dissipation, (5) surface smoothness and coating hardness when the part requires a mating surface, (6) compatibility with various cleaning methods, (7) coverage of intermetallic compounds, (8) outgassing characteristics, and (9) cost.

3.1 Chromate Conversion Coating

The chromate conversion coating process can be used on aluminum, magnesium, and steel substrates. When a metallic part is dipped in a chromating solution, the metal surface layers are dissolved, and a thin film, approximately 10 nm (0.394 μin.) thick, of chromium chromate and metallic chromate along with a few molecules of hydration is formed. When the coating is scratched or damaged, the hexavalent chromium complex, with the help of the water of hydration, seeps into the damaged area and repairs the coating. This self-healing characteristic is the primary advantage of chromate conversion coatings.

Because the coating is thin, uniform, and dip-coated, it is well-suited for parts with complex shapes and tight dimensional tolerances. Chromate conversion coatings cannot form on intermetallics, and its discontinuity at those locations could lead to atmospheric exposure of the substrate, with subsequent corrosion. Figure 3.1 shows the discontinuity of a chromate conversion coating, caused by an intermetallic on an aluminum alloy surface, at a defect site.

Because the coating is thin, it does not affect the electrical conductivity of the substrate and does not change the surface finish. It does not exhibit any outgassing problems. Although chromate conversion coatings are inexpensive, the cost of disposal of the spent solutions containing hexavalent chromium ions could be quite high due to environmental regulations, because these hexavalent chromium complexes are proven carcinogens.

The coating is a continuous layer with no porosity and provides good molecular adhesion with organic layers. Consequently, it is often used as an undercoat for organic coatings. Chromate conversion coatings are often used as an underlayer, when polished aluminum substrates are coated with epoxy-based particulate gamma iron oxide magnetic paint in the manufacture of rigid disks.

Chromate conversion coatings lose many of their useful properties when heated in air at temperatures above 70 °C (160 °F) for extended periods. One such property is its self-healing capability. At elevated

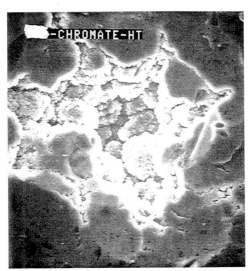

Fig. 3.1 SEM micrograph of an aluminum alloy surface with defects caused by intermetallics. The chromate conversion coating is discontinuous at the defect sites.

temperatures, the chromate loses its water of hydration, and the hexavalent chromium transforms into trivalent chromium, which prevents the chromium from moving on the surface to repair the damaged coating.

This type of coating slowly dissolves in water; consequently, prolonged exposure to water damages it. Also, the hexavalent chromium ions are leached into water in which the part is immersed. Because of the environmental concerns about chlorofluorocarbons (CFC), alternate cleaning techniques are being pursued, the most promising being aqueous cleaning, either in jet form or with ultrasonics. When hexavalent chromium leaches into cleaning water, disposing of the water becomes an environmental concern. As a result, chromate conversion coating is becoming less popular as a corrosion-resistant surface finishing operation.

Magnesium alloys are used increasingly in the disk drive industry because of their higher strength-to-weight ratio than aluminum and zinc alloys. Dow Chemical Company has developed a special chromate coating called Dow-7 for magnesium alloys. Properly applied, Dow-7 varies in color from light to dark brown. It causes no appreciable change in dimensions. Prior to Dow-7 application, the magnesium surface is cleaned and activated in an acidic fluoride pickling bath containing an aqueous solution of sodium, potassium, and ammonium fluorides.

3.2 Anodizing

Anodizing is useful for aluminum and magnesium substrates. The part to be coated is made the anode in an electrolytic cell with a suitable electrolyte such as sulfuric, chromic, or oxalic acid. The oxygen that is generated at the anode during electrolysis oxidizes the metal. A thin oxide layer on the metal should stop further oxidation under normal circumstances. However, during anodizing, the oxide layer partially dissolves in the electrolyte, allowing more metal to be exposed to oxygen, thereby enabling the oxide layer to grow thicker. Typically, half of the layer grows inward and the other half outward of the original metal-fluid interface.

The plating parameters can be modified to obtain a pore-free hardcoat (for higher hardness and lower particle generation) or a porous, conventional anodized layer (for coloring and sealing). For disk drive applications, because contamination is of primary importance, the pore-free hardcoat is always preferred. See Case Study 3.1

for a discussion of how an anodized layer could generate particulate contamination.

Anodizing is used to increase the wear resistance of a part when it is used in a load-bearing application. The coating can also be used to coat the surface with a thick oxide layer for corrosion resistance. Anodizing is well suited for parts with simple, flat shapes. For a complex shape with blind corners, the uniformity of the coating is not satisfactory because of the nonuniform current distribution at the edges and corners during the anodizing process. Too much variation in current can lead to "burn marks," which consist of loose and powdery oxide particles that can easily become airborne.

Anodized layers act as insulators, and an anodized wire can be wound into a coil without shorting. However, for many parts, the insulating layer can cause charging problems. If the conductivity of the anodized layer can be improved by some innovative process change, this type of coating will find many more applications in the disk drive industry.

Outgassing is not a problem with anodized coatings. During aqueous cleaning, water reacts with the top layer, forming a soft and spongy layer of pseudo-boehmite. Addition of 5 to 10% of ortho-phosphoric acid to the rinsing water inhibits the boehmite formation (Ref 1).

Because the coating is thick, coverage of intermetallics is satisfactory. For magnesium alloys, Dow-17 is the proprietary anodizing treatment. It is a two-phase, two-layer coating. A light green undercoating of approximately 5 μm (197 μin.) forms at lower voltages. A much heavier, second-phase coating, approximately 30 μm (1180 μin.) thick of dark green color covers the underlayer. The top layer is brittle, but has good abrasion and corrosion resistance. Its corrosion resistance can be increased by dichromate-bifluoride post-treatment (part of a proprietary process called HAE treatment).

3.3 Passivation

Passivation is an inexpensive chemical immersion technique, specially tailored to stainless steel, which owes its corrosion resistance to the formation of a thin, continuous, and adherent layer of chromium oxide (chromia). If the piece is subsequently machined, some of the tool steel debris becomes stuck on the surface and corrodes in the presence of moisture to form brown stains of iron oxide. There is no protective chromia in those areas, and corrosion could proceed unhindered.

As a final processing step, if the stainless steel part is immersed in dilute (approximately 10%) nitric acid, the acid will react and remove all iron atoms from the surface and near-surface layers. This leads to the formation of a continuous, adherent chromia layer. Because chromia has higher hardness than steel, passivated steel has higher microhardness. Over-passivation makes the chromia layer thicker and the surface more brittle.

Passivation removes many intermetallics effectively. The passivation film does not outgas and is compatible with any surface cleaning technique. It is very thin (on the order of a nanometer), and it does not affect either electrical conductivity or surface finish. Because passivation is an immersion technique, the reaction is quite uniform even on complex-shaped parts, and the dimensional specifications of the part are maintained.

3.4 Organic Coatings

Organic coatings, either electrostatic spray coated or electrostatic dip coated, inhibit generation of particulate contamination on the surfaces of a variety of materials. This technique is very versatile and inexpensive. Chromate conversion coatings provide strong molecular adhesion of the coating with the substrate and thus are used as undercoats for organic coatings.

These coatings are generally 20 μm (787 μin.) or more thick. Their thickness cannot be controlled very well from one location to another on the part, or from part to part. As a result, organic coatings cannot be used on parts with tight dimensional tolerances. They are soft and have poor wear resistance and consequently cannot be used for parts with mating surfaces. These coatings exhibit poor scratch resistance and can be easily damaged during handling and assembly. Figure 3.2 shows the damage and loose contamination generated by a handling scratch on an organic coating.

These coatings are generally electrical insulators. It is often necessary to machine the coating off the surface of the part to provide a mating surface and/or to provide electrical grounding. The machined area needs to be protected with a chromate conversion coating. The edges of the machined area are sources of flaking of the coating and are potential contamination generators. The loose contamination

Fig. 3.2 Handling damage on the surface of the soft organic coating on a metal substrate. The scratch can generate loose particulate contamination.

generated at the edge of the machined surface is the most significant problem with organic coatings.

Figure 3.3 shows breaking and flaking of the relatively brittle electrostatic coating at the edge of a machined surface. Figure 3.4 shows the loose contamination present when the electrostatic coating flakes. It is very expensive to mask the required area and then apply the organic coating.

Good adhesion of the coating to the substrate requires thorough and immaculate cleaning of the surface. Generally, the edges of castings are covered with thicker oxide scales and other contaminants and need more extensive surface preparation. Fig. 3.5 shows a magnesium casting coated with an electrostatic coating, in which the edges were not coated due to inadequate surface preparation. If the surface contains impurities even after cleaning, the electrostatic coating exhibits defects, as shown in Figure 3.6, in which an island of uncoated metal is visible at the center.

Many organic coatings allow permeation of moisture to some extent. If moisture penetrates through the coating or a coating defect, water reacts with the substrate material, forming coating bubbles and potential contamination. See, for example, Case Study 3.3.

Fig. 3.3 Chipping and flaking at the edge of the machined area of an electrostatic-coated metal substrate. The coating is brittle and prone to chipping.

Fig. 3.4 Loose particulate contamination generated at the edge of the machined area due to chipping of the brittle electrostatic coating.

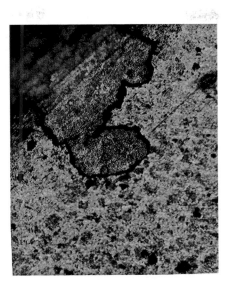

Fig. 3.5 Magnesium alloy casting coated with an electrostatic coating. At a sharp area of a casting, the surface has more impurities and mill scale. Due to inadequate surface preparation, the edges were not coated. 600'.

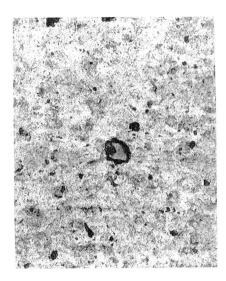

Fig. 3.6 Missing island at the center where the electrostatic coating did not cover, caused by surface impurity on the metal substrate. 200'.

Outgassing is a potential problem for some coatings. With an organic coating, the surface is quite homogeneous, and the coverage of intermetallics is excellent. However, the surface should be free of mill scale, hydrocarbons, and other contaminants so as to promote good adhesion of the coating.

3.4.1 *Conformal Coatings*

Conformal cold coatings (e.g., Parylene developed by Union Carbide) are considered organic coatings. Parylene C, or poly chloro-*p*-xylylene, is a linear, high molecular weight polymer deposited *in vacuo* directly from the gas phase. The dimer, a crystalline solid, is heated in vacuum. The evaporated dimer is pyrolyzed at 600 to 700 °C (1110 to 1290 °F). The gaseous monomer enters the deposition chamber where it polymerizes on to the part, which is kept at room temperature.

Parylene has very good corrosion resistance because it is a polymer. It is approximately 2 μm (79 μin.) thick and provides excellent coverage on complex-shaped parts. It is insulating and has a smooth surface, but exhibits low wear and scratch resistance. This coating does not dissolve in water and common organic solvents; consequently, it is compatible with most cleaning techniques. It provides good coverage of intermetallics, does not experience any outgassing problems, and is relatively inexpensive.

3.5 Electroless Nickel

With appropriate phosphorus content, electroless nickel (EN) can be nonmagnetic, amorphous, hard, and conducting. It forms a uniform, thin layer even on complex shapes. Its thickness uniformity and control are excellent. Electroless nickel is widely used on aluminum, magnesium, steel, and any other substrate for corrosion and wear resistance, contamination control, and EMI/RFI shielding. See Appendix 3.1 for various electroless nickel applications.

For electroless nickel to adhere strongly to the substrate, the surface should be cleaned thoroughly. Intermetallics should be removed because the coating does not adhere well in their presence. Chemical etching is a convenient way of cleaning the surface of surface contamination, mill scale, and intermetallics. Experience indicates that very harsh chemical cleaning weakens the surface metal grains and intermetallics, which could lead to particle generation. Mechanical cleaning techniques such as tumbling and bead blasting will not weaken the surface and, moreover, they introduce beneficial compressive stresses into the surface. Appendix 3.2 discusses the use of electroless nickel coatings on cast aluminum alloys, and Appendix 3.3 discusses their use on cast magnesium alloys.

3.6 Ion Vapor Deposited Aluminum

Ion vapor deposition of aluminum is a dry, nonaqueous technique of depositing pure aluminum to a specified thickness in an evacuated chamber. Aluminum is thermally evaporated, energized with a plasma, and deposited on the part. This technique has been developed for aerospace applications and is slowly finding its way into disk drive manufacture.

Ion vapor deposited aluminum acts as a sacrificial anode to the substrate and improves the stress corrosion and fatigue resistance of the substrate. It is a continuous layer and provides excellent electrical contact and electromagnetic interference compatibility.

The ion vapor deposited layer can be surface finished in the same way as aluminum. For example, it can be chromate conversion coated (Ref 2). It can be deposited without a significant increase in the temperature of the substrate. Because it is a dry technique, it does not experience outgassing or fluid entrapment.

Ion vapor deposition can cover up intermetallics very effectively. Because it is a pure aluminum layer, any surface coating will adhere very well to that layer. The chromate conversion coating on ion vapor deposited aluminum is uniform and pore-free (see Fig. 3.7) and has excellent adhesion.

3.7 Plasma Polymerization Coating

Plasma processing is a surface modification treatment using ionized gases at temperatures slightly above room temperature. It enhances the surface properties without affecting the bulk properties.

In plasma polymerization, ultra-thin coatings are deposited by introducing carbon-containing gases into the active plasma. The carbon in the gas forms the backbone of a long-chain polymer that is deposited on the surface of the part.

Fig. 3.7 Uniform, pore-free chromate conversion coating over ion vapor-deposited aluminum coating on the aluminum alloy casting. Intermetallics and other surface defects are covered effectively by this ion vapor coating.

Plasma polymerization coatings are defect-free and adhere strongly to all substrates. They are chemically very inert and exhibit excellent corrosion resistance. They are well-suited to all materials. Plasma polymerization is a clean, dry process and does not encounter any waste disposal problems.

The coatings are smooth and can be very thin, 20 nm (0.79 μin.) to several micrometers, so that the part will have excellent dimensional tolerance. They have excellent uniformity even on complex-shaped parts. Although they are not conducting, these coatings are quite thin and have sufficient conductivity for many applications.

They adhere tenaciously to the substrate, and they have good lubricity. However, their wear resistance is low. Because they are chemically inert, they are compatible with most of the cleaning techniques. Plasma polymerized coatings provide excellent coverage of intermetallic compounds of the substrate.

They do not outgas, because they are polymerized. Initial setup costs are high, and surface preparation for the coating must be meticulous.

3.8 Salt Bath Nitriding

Salt bath nitriding (Ref 3) is used for ferrous metals only. Although its primary application is to improve wear and fatigue properties, it also enhances corrosion resistance. It is a thermochemical diffusion process, in which the ferrous part is immersed in salts with a particular nitrogen potential. Nitrogen reacts with iron to form a tough, ductile, and corrosion-resistant compound called epsilon iron nitride (Fe_3N).

The nitrided layer can be made as thick as 1 mm (0.039 in.). Kolene Corporation has developed proprietary modifications of this process, called Tuffride and Melonite. It exhibits good corrosion resistance and controlled thickness. Salt bath nitriding provides excellent coating uniformity on complex-shaped parts. It has superior hardness and satisfactory electrical conductivity and is compatible with all cleaning techniques. No outgassing is experienced.

Its intermetallic coverage is suspect. It can be used only on ferrous parts and involves temperatures of 700 °C (1290 °F). Additionally, it is a relatively expensive process.

3.9 Case Histories

CASE STUDY 3.1: Porous Surface of Anodized Hardcoat

Problem. Linear actuator bearings, made of steel, were designed to move over an aluminum alloy rail. To provide the required wear resistance for aluminum, the part was anodized to obtain a hardcoat using sulfuric acid anodizing. During testing of the bearing combination, significant particulate generation from the anodized layer was observed.

Analysis. The hardness of the two mating parts was checked and found to be according to specification. The thickness of the anodized layer also did not change. Consequently, the anodizing process was found to be under control. However, the top surface was found to have very fine pores, making the top layers brittle.

Solution. About 30% of the anodized layer was lapped away before the bearing combination was tested. No particulate contamination was generated.

Recommendation. Even though the anodized layer was supposed to be a pore-free hardcoat, sulfuric acid anodizing provides an anodized layer with fine pores in the top layers. The pores make the layers brittle and generate contamination in a load-bearing application. Because sulfuric acid is a convenient electrolyte, a decision was made to use the same anodizing process, but to remove the top 30% of the layer by lapping to provide a pore-free bearing surface.

CASE STUDY 3.2: Filiform Corrosion Under an Electrostatic Coating

Problem. A magnesium alloy part coated with an electrostatic coating exhibited growth of a powdery substance (in the form of a snake) beneath the coating, leading to the formation of a bubble in the coating. The part was exposed to normal processing conditions.

Analysis. The part was cross sectioned across one of the bubble defects. Figure 3.8 shows that the coating lifted and formed a bubble due to generation of a corrosion product. Figure 3.9 shows the shiny corrosion product on the metallic substrate, and Fig. 3.10 shows the penetration of the corrosion product into the metal substrate through the surface porosity.

Solution. Filiform corrosion occurs when fine solid contaminants are present on the metal under the coating. The coating is weak and

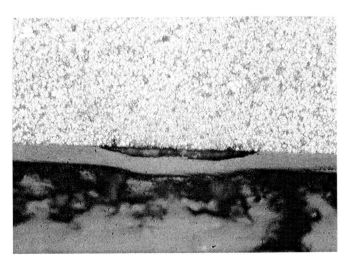

Fig. 3.8 Lifting of the electrostatic coating and bubble formation due to corrosion on the magnesium alloy surface. Cross section of the defective area is shown. 200'.

Fig. 3.9 Shiny corrosion product on the metal surface which caused the coating to lift. Corrosion was caused by moisture penetration through the defective coating on the surface contaminant. 600'.

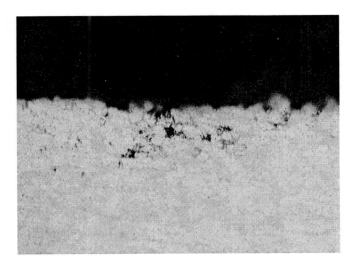

Fig. 3.10 Oxidation proceeding into the metal due to surface porosity. Entrapment of particulate contamination allowed moisture penetration through the coating. 600'.

porous in that area, and moisture can penetrate the defective coating. If the contaminant has an active species such as chlorine, the metal and the metallic chloride become oxidized in the presence of moisture, thereby generating active chlorine for further corrosion. Generally, chlorine is concentrated in the head of the corrosion trail. Corrosion proceeded into the metal through the surface porosity.

Recommendation. Some of the contaminants that can cause filiform corrosion are industrial waters and seawater (e.g., NaCl, $CaSO_4$, and $MgSO_4$), processing chemicals (e.g., chromic acid), dust particles (e.g., silicates), and organic material. The metal surface should be cleaned well and tested by turbidity testing, for example, before it is coated with the electrostatic coating.

CASE STUDY 3.3: Dow-7 and Poor Adhesive Bonding

Problem. A component made of AZ-91B magnesium alloy was coated with Dow-7 coating (chromate conversion coating used for corrosion protection of magnesium) and then bonded to a flex circuit with an adhesive. One particular lot was found to have very poor peel strength.

Analysis. Optical microscopy revealed that during peeling the failure occurred at the interface of the metal-Dow-7 coating. Several metal samples from the acceptable and unacceptable lots were cross sectioned. Figure 3.11 shows a cross section of the magnesium alloy component with the Dow-7 coating, taken from an unacceptable lot that had poor peel strength. The coating was cracked. Figure 3.12 shows a cross section of a component from the acceptable lot with excellent peel strength. The coating was crack-free. The coating thickness of the failed lot was 1.07 to 1.52 μm, (42 to 60 μin.) whereas it ranged from 0.76 to 1.10 μm (30 to 43 μin.) in the acceptable lot.

Solution. Because the coating was cracked, it came off with the flex circuit adhesive at low peel strengths. The cracks were probably caused by mismatch in thermal expansion between the metallic substrate (with much higher thermal expansion coefficient) and the coating. As the coating increases in thickness, it becomes more brittle, and the thermal mismatch will be more pronounced.

Recommendation. The coating thickness was reduced. No quenching or sudden thermal shock was allowed to occur during processing.

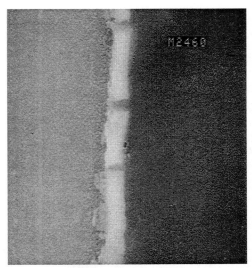

Fig. 3.11 Cross section of a magnesium alloy component with the white Dow-7 coating, from the unacceptable lot. The coating cracked because it was thicker and was exposed to a thermal shock. The cracked coating peeled off easily.

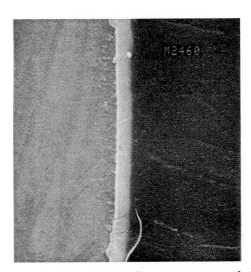

Fig. 3.12 Cross section of magnesium alloy component with the white Dow-7 coating, from the acceptable lot. The coating was thinner and crack-free, and it had excellent peel strength in an adhesive bond with a flex circuit.

CASE STUDY 3.4: Bubbling of Powder Coating Caused by Moisture

Problem. A magnesium alloy part was coated with a powder organic coating and placed in a disk drive. After running the drive for 6 months, it was opened for part evaluation. The magnesium part displayed bubbles on the surface. Poor adhesion to the surface was suspected.

Analysis. Paint adhesion tests did not show any abnormally low adhesion strength. The bubbles were studied in a scanning electron microscope (SEM). Figure 3.13 shows a bubble in the coating with a pinhole at the center.

Solution. The coating had a pinhole defect, probably caused by the presence of a particulate contaminant on the surface during the coating process. Moisture was trapped in the pinhole during aqueous cleaning. Water reacted with magnesium forming magnesium oxide, which pushed the coating up to form a bubble. The coating had sufficient flexibility to accommodate the bulge and was satisfactory.

Recommendation. The surface should be free of contamination during coating.

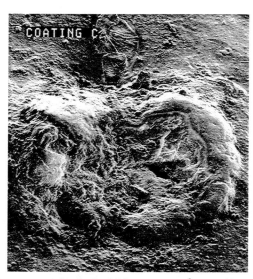

Fig. 3.13 SEM micrograph of a bubble in a powder coating on a magnesium alloy. The pinhole in the coating allowed moisture entrapment and consequent corrosion of the magnesium substrate.

CASE STUDY 3.5: Bubbles and Pinholes Caused by Substrate Defects

Problem. An aluminum die cast part coated with an electrostatic coating showed bubbles and pinhole defects in the coating.

Analysis. Figure 3.14 is a SEM micrograph of a defect on the coating showing a pinhole at the center of the defect. At the center of the pinholes, aluminum alloy intermetallics were noticed. Figure 3.15 shows bubbles that were present in the coating. Under the bubbles, the casting exhibited fine pores.

Solution. When intermetallics are present on the alloy surface, they are not coated satisfactorily. In those sites, moisture penetrates through the coating, reacts with the metal, and forms a bubble around the pinhole.

Casting surface pores trap moisture during machining and pre-cleaning before the electrostatic coating covers them. As the metal reacts with the trapped moisture, forming oxide, the coating bulges, forming bubbles.

Fig. 3.14 Bubble with a pinhole at the center found on the electrostatic coating over an aluminum alloy cast part. Pinholes formed due to substrate surface defects such as intermetallics and moisture penetration through the coating.

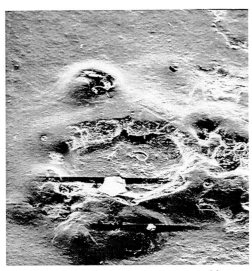

Fig. 3.15 Bubbles in an electrostatic coating caused by moisture trapped in the surface pores of the casting.

Recommendation. Higher purity alloys have fewer intermetallics and should be used. The casting procedure should be improved (e.g., high-pressure casting) to reduce surface porosity.

CASE STUDY 3.6: Coating Bubbles Caused by Organics

Problem. An iron part was made by powder metallurgy and then impregnated with an organic sealant to seal the surface porosity. The part was then machined and coated with an electrostatic coating. Bumps and other defects were noticed on the coating.

Analysis. Figure 3.16 shows the cross section of a bump defect in the coating. The substrate under the defect has a shallow hole visible. Figure 3.17 shows the metal surface with porosity under a coating defect, and Fig. 3.18 shows organic material under the defect.

Solution. The sealant used for impregnation was unstable and caused defects. The powder metallurgy part was polished, etched, and studied. It was determined that the part was quite dense and did not require sealant.

Fig. 3.16 Bump in the electrostatic coating over a shallow hole on the surface of a powder metallurgy part. 50'.

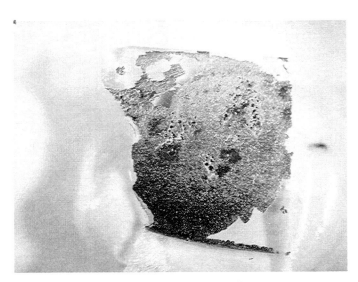

Fig. 3.17 Surface porosity of the powder metallurgy part beneath a bubble. 200'.

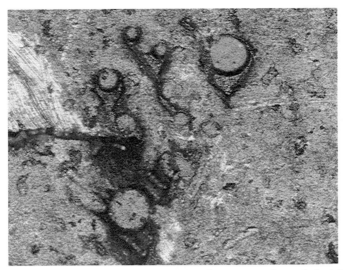

Fig. 3.18 Sealant material on the metal substrate under a defect. The sealant operation contributed to bubble formation and could be eliminated. 500'.

Recommendation. The impregnating sealant operation can be eliminated. After machining, the surface should be cleaned well before application of the coating.

References

1. A.S. Brar and P.B. Narayan, US patent pending
2. A.S. Brar and P.B. Narayan, US patent 5,017,439, May 21, 1991
3. E. Taylor, QPQ: Salt Treatment that Prevents Corrosion, *Metal Prog.*, July, 1983

Appendix 3.1

Electroless Nickel for Computer Disk Drives*

As computer technology advances, computer disk drives must store more and more data. Therefore, the specifications for components of disk drives and other computer peripherals are becoming ever more exacting. Electroless nickel (EN) plating is already playing a part in meeting these specifications, but some problems remain to be solved to expand its use in the future.

The four most important features of electroless nickel that make it an excellent finish for disk drives (Fig. 1) and other computer peripherals are its nonmagnetic characteristics, its adhesion to almost all substrate materials, its hardness, and its uniform coverage of parts of intricate geometry.

Electroless Nickel Plating

The electroless nickel plating bath discovered by Brenner and Riddel (Ref 1) consists of three major components—a nickel salt, a reducing agent (e.g., sodium hypophosphite) that decomposes the nickel salts into nickel ions, and a buffer (e.g., hydroxy acetic, glyceric, citric or succinic acid), which also acts as a mild complexing agent for nickel. There are other additives that improve the properties and deposition rates. The result is the deposition of a nickel-phosphorus or nickel-boron alloy.

There are basically two types of baths: acidic (pH 4 to 7) and ammoniacal (pH 8 to 11). The hot acidic baths are used to plate steel,

---

* S. Kaja, P.B. Narayan, and A.S. Brar, *Products Finishing*, Feb, 1988, p 46-56. Reprinted with permission.

aluminum, and other metals, whereas the warm alkaline baths are used to plate plastics and nonmetals. A typical acidic bath formulation is shown in Table 1.

Table 1   Typical Acid EN Solution

Nickel chloride ($NiCl_2.5H_2O$)	20 to 30 g/L
Nickel sulfate ($NiSO_4.6H_2O$)	0 to 25 g/L
Sodium hypophosphite ($NaH_2PO_2.H_2$)	10 to 25 g/L
Hydroxy acetic acid ($HOCH_2COOH$)	0 to 30 g/L
Sodium citrate ($Na_3C_6H_5O_7.2H_2O$)	0 to 12 g/L
Sodium acetate (CH_3COONa)	0 to 10 g/L
Sodium hydroxide ($NaOH$)	to neutralize
pH	4 to 7 g/L
Temperature	90 to 100 °C (194 to 212 °F)

Fig. 1   Cross section of a hard disk drive.

Fig. 2 Disk drive with electroless nickel-plated components.

Microstructure

Although electroless nickel has been a popular coating on metals and nonmetals for nearly 40 years, there are few reported studies regarding its microstructure, crystallinity, and the mechanism by which deposits are hardened.

Researchers using X-ray diffraction and optical microscopy have found an amorphous structure. However, a study by Graham and co-workers (Ref 2), using transmission electron microscopy, showed that electroless nickel can have a very fine crystalline structure and that, in the as-plated condition, it is a super-saturated solid solution of phosphorus in crystalline nickel. Later investigations revealed that electroless nickel is crystalline up to a certain phosphorus content, typically about 9 wt%, becoming amorphous at higher phosphorus contents.

Physical Properties

The pH of an electroless nickel solution, its temperature, and composition play important roles in determining the physical prop-

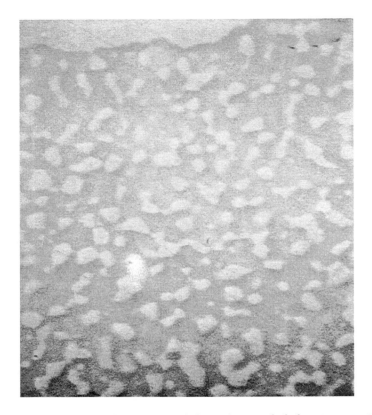

Fig. 3 Microstructure of heat treated electroless nickel showing precipitation of nickel-rich phase (white islands).

erties of deposits, while also influencing the deposition rate. These are important determinants of physical properties such as hardness and smoothness.

Hardness of as-plated electroless nickel varies from 250 to 700 HV. Hardness is generally lower at high phosphorus content. The strain in the nickel lattice due to the presence of phosphorus resists plastic deformation and hence increases hardness. At high phosphorus content, the nickel alloy loses its crystallinity, thereby reducing the hardness of the deposit.

Heat treatment increases hardness. It separates the alloys into a solid solution of phosphorus in nickel and an intermetallic com-

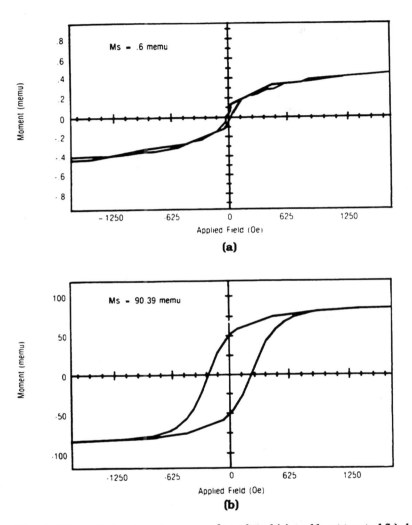

Fig. 4 Magnetic hysteresis curves of as-plated (a) and heat treated (b) electroless nickel, indicating nonmagnetic transformation as a result of heat treating. The saturation magnetization increased from near zero to 90 memu during the heat treatment. Notice the change in scale for the ordinate.

pound of nickel and phosphorus (usually Ni_3P) (Fig. 3). Hardness increases as a result of precipitation hardening.

Magnetism. As-plated electroless nickel deposits are magnetic at very low phosphorus contents and nonmagnetic at higher phosphorus contents. Apparently, phosphorus disturbs the electronic coupling in nickel, making the deposit nonmagnetic.

When the phase separation occurs during heat treatment, the nonmagnetic deposit becomes partially magnetic because the solid solution phase is magnetic. As expected, deposits with lower phosphorus contents become magnetic with milder heat treatments. Figure 4 shows a hysteresis curve from deposits in as-plated and heat treated (at 400 °C, or 750 °F, for 1 h) form. It shows the nonmagnetic-to-magnetic transformation of electroless nickel occurring during heat treatment.

Corrosion Resistance. Electroless nickel has excellent corrosion resistance. It is relatively unaffected by acidic and basic environments. The exact mechanism of chemical resistance is not well understood, but it can be assumed that phosphorus ties up the chemically active electrons of nickel.

Adhesion. For disk drive applications, the most important property is adhesion. As in any plating process, the substrate must be active when exposed to the plating solution. It is important to eliminate the smut and other contamination from the surface to improve adhesion and provide a strong metallic bond with the electroless nickel deposit.

Electroless Nickel Testing

Electroless nickel deposits are commonly tested for hardness, smoothness, brightness, and adhesion. For many high-tech applications, however, the electroless nickel deposit is as thin as a micrometer, and thus, the conventional mechanical testing procedures cannot determine the true properties. There is a need to develop new techniques to study micromechanical properties such as hardness, toughness, and internal stress of very thin films.

There is an additional problem in studying adhesion. The tests are very complex and usually require plating a special test coupon. Because most the parts coated with electroless nickel have intricate shapes and varying surface finish requirements, the adhesion test results obtained from a test coupon have limited value.

There is a need for a good quantitative test that can be used on production-type samples. All of the available adhesion tests that can be used directly on the product are semiquantitative at best. The most commonly used adhesion test is the tape test, which is qualitative but can be made more quantitative with the help of a densitometer.

Disk Drive Applications

Underlayer. Electroless nickel is being used as an undercoat for magnetic media applied on the surfaces of aluminum disks. It serves several purposes.

Aluminum alloy 5086 is commonly used as a substrate for thin-film rigid disks in magnetic storage disk drives. However, when a magnetic layer of CoNiCr/Cr is sputtered onto this alloy, a galvanic couple between the magnetic layer and the aluminum can form, accelerating corrosion. In addition, aluminum alloys tend to include intermetallic compounds (i.e., AlFeSiMn) that do not properly accept the deposition of the magnetic layer. Thus, the intermetallics show up as errors and missing bits during recording. Intermetallics also can trap moisture and contaminants that could initiate corrosion. An electroless nickel deposit under the magnetic media prevents these problems.

Another purpose of electroless nickel in this application is to provide a hard surface for deposition of the magnetic media. Electroless nickel under the magnetic film enhances the wear and tribological properties of the head/disk interface.

Because no other magnetic material can be tolerated under the magnetic layer, the electroless nickel deposit applied as an underlayer must be able to resist magnetic transformation at temperatures of 200 to 300 °C (390 to 570 °F) during sputter deposition of magnetic media.

Overcoat. Electroless nickel is a tough coating with many of the characteristics required to protect the surface of a magnetic disk. The interaction of the hard magnetic disk and the read/write head is technologically the most complex part of the drive. Extremely tight tolerances are required.

The disk must have a surface roughness less than 0.02 μm. The metallic magnetic medium is about 0.05 μm thick.

In the recent Winchester technology, the head takes off and lands on the disk. The core of the head, surrounded by ceramic, lifts off the surface of the disk as the disk rotational speed increases. It then "flies" 0.25 μm above the disk rotating at 3600 rpm.

If there is vibration in the disk drive, minute irregularities on the surface of the disk, or any other cause of instability in the head flying on its air bearing, the head can strike the disk while the disk is still rotating at 3600 rpm. This could cause "head crash," which generates enormous stresses at the head/disk interface and may remove portions of the vital magnetic media from the surface of the disk.

Electroless nickel is being touted as the optimum overcoat on the aluminum alloy substrate because of its excellent coverage of intermetallics, its resistance to corrosion, its high hardness, and good tribological characteristics.

Microcontamination. Because of the extremely small head/disk tolerances, microcontamination from any source can have a devastating effect on the head/disk interface, possibly leading to head crash. Many overcoats are being considered for use on the various components, but electroless nickel has a very good change of becoming the optimum coating. It helps to avoid microcontamination.

Powder Metallurgy. There is an aluminum drive component requiring extremely small dimensional tolerance (about 10 µm). It is often made of wrought aluminum followed by very expensive precision machining. A less expensive alternative would be a part made by powder metallurgy. The surface of the sintered material is porous, however, and can absorb unwanted species or generate microcontamination. If the sintered material is coated with electroless nickel, the porosity is covered very effectively (Fig. 5).

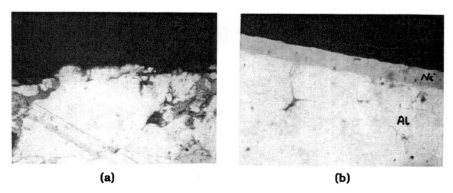

(a)   (b)

Fig. 5 Sintered aluminum alloy (a) exhibits significant surface porosity that could lead to absorption of contaminants. When the part is electroless nickel plated (b), the porosity is well covered.

Steel. Electroless nickel prevents corrosion of various steel parts. As an example consider an end plate that closes the magnetic flux lines for the magnet assembly. It had been zinc plated, but exhibited corrosion during humidity testing. It was a potential generator of loose particles. When the zinc plate was replaced with a thin film of electroless nickel, the end plate passed the humidity tests.

Brake Drums. The brake drums in disk drives have been made of plain carbon steel, coated with electroless nickel. The drums from one lot wore brake pads very quickly, however. An SEM examination found that the bad lot had nodules that caused excessive wear on the pads (Fig. 6). An acceptable lot had a much smoother surface (Fig. 7). The deposition conditions were changed so as to obtain a smooth surface, and the brake pad wear problem was solved.

Diffusion Barrier. Electroless nickel serves as a diffusion barrier between copper and gold deposits on printed circuit boards. It prevents the interdiffusion of these two layers. Printed circuit boards are usually plated with copper for good electrical and thermal conductance. Copper oxidizes easily, however, thus increasing contact resistance. Gold is generally plated on copper to prevent oxidation. However, copper can readily diffuse through gold and come to the

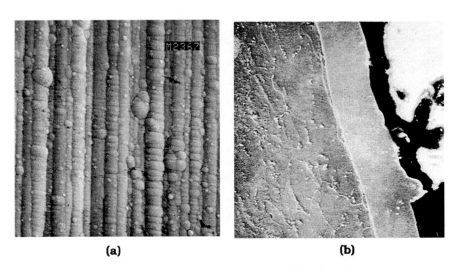

(a)   (b)

Fig. 6 Electroless nickel deposit with sharp nodules (a). This deposit, applied on a brake drum, caused rapid wear of brake pads. Cross section of brake drum with electroless nickel plating (b).

surface to oxidize, thus again increasing contact resistance. An intermediate layer of electroless nickel prevents this.

In less critical applications, where electrical contact and corrosion resistance of copper are not extremely important (e.g., when the circuits are covered with an epoxy), electroless nickel is used to coat the copper, eliminating the use of gold.

In this application, electroless nickel provides not only excellent wear resistance, but also good solderability. Copper is difficult to solder because it oxidizes easily.

EMI/RFI Shielding. The delicate electromagnetic interactions at the head/disk interface in a disk drive should be shielded from external EMI/RFI (electromagnetic interference/radio-frequency in-

Fig. 7 Plating conditions were changed to provide smoother deposit for brake drum shown in Fig. 6.

terference). A deposit of electroless copper provides excellent shielding. Although a copper deposit provides the required shielding, it has poor resistance to oxidation and wear, and thus long-term protection from EMI/RFI may suffer. An electroless nickel overcoat is almost invariably required. When lower levels of shielding or simple protection from electrostatic discharge is required, the electroless nickel deposit alone may be sufficient.

Lubricity. Most of the components in a disk drive are made of aluminum alloys and stainless steels. When stainless steel screws are driven into and taken out of an aluminum alloy part, there is a significant amount of gouging and material removal. However, when a thin layer of electroless nickel is deposited on the screws, the gouging problem disappears, apparently because of the lubricity of electroless nickel. The presence of phosphorus can be assumed to cause the amorphous nickel layers to readily slide one over the other, providing effective solid lubrication. Figure 8 shows other components that are being considered for electroless nickel plating.

Electroless Nickel Adhesion

Many disk drive components are made of 380 die cast and 356 sand cast aluminum. Coating with electroless nickel minimizes shedding of microcontaminants and improves corrosion resistance.

However, adhesion of electroless nickel on cast aluminum can be a problem. Too often, tape tests for adhesion result in an unacceptably large number of metallic particles being pulled from a cast surface. The most probable culprit is smut and other contamination

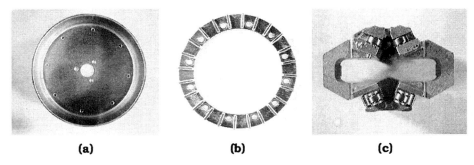

(a)   (b)   (c)

Fig. 8 Electroless nickel plating is being considered for these components of computer disk drives.

present on the as-cast surfaces and not completely removed by routine cleaning and pretreatment.

The casting alloys have considerably higher silicon content than the wrought alloys because silicon increases the fluidity of the molten metal. Because silicon oxide scale is very adherent, however, it cannot be desmutted easily. The adhesion problem seems to be worse on sand castings than on die castings.

Electroless nickel adheres better on wrought alloy surfaces than on cast surfaces. The latter has a significant amount of smut, mold release, and other contamination and therefore have to be strongly etched to remove the contamination. During etching, the surface becomes very rough, and as a result, a thin electroless nickel deposit can easily fracture at high and low points on the surface, thus reducing its adhesion to the casting.

When the cast surface is machined, a light etching will be sufficient to prepare the surface for electroless nickel plating. Thus, the electroless nickel has much better adhesion to the smoother machined surface than to the cast surface.

In tape testing for adhesion of electroless nickel on aluminum alloy castings, the loose particles removed from the surface could come from lack of adhesion at the electroless nickel/substrate interface and/or from lack of adhesion of the electroless nickel deposit to itself. More study is needed to understand the adhesion of electroless nickel to the as-cast surface and how it could be improved.

Experience shows that electroless nickel provides a clean and contamination-free coating on aluminum alloy surfaces, but its adhesion on as-cast surfaces is not up to the stringent requirements of contamination control in disk drives.

Other Problems

Plating Control. One of the most difficult problems in electroless nickel plating is consistently controlling the plating parameters such as composition, bath uniformity, and temperature. Taking the cue from the philosophy of Dr. Genichi Taguchi regarding product manufacturing, more effort is needed to provide bath compositions that provide deposits with properties not unduly affected by the plating variables.

Iron Contamination. Particularly when coating steel, iron enters the plating bath and redeposits as iron particles in the electroless nickel deposit. These particles provide active nucleation sites for electroless nickel, causing the growth of nickel nodules, which result in a rough surface. Modifications of the bath to capture the iron and eliminate the roughness are needed.

Microcontamination. More work is needed to eliminate loose particles and microcontamination from the surface to be plated. As the use of electroless nickel becomes more popular for high-tech components with tight dimensional tolerances, keeping microcontamination out of the electroless nickel deposit will be more and more important.

As lighter magnesium and zinc die cast alloys get ready to replace aluminum alloys in high-performance disk drives, the role of electroless nickel as a protective overcoat for the corrosion-prone magnesium and zinc alloys will be much more crucial.

Electroless nickel platers should tackle aluminum alloy castings now and get ready for magnesium and zinc alloy castings.

References

1.   A. Brenner and G.E. Riddell, *J. Res. Nat. Bur. Stand.*, Vol 37 (No. 1), 1946; *Proc. Am. Electroplate. Soc.*, Vol 33 (No. 16), 1946; US patent 2,532,284, Dec 5, 1950
2.   A.H. Graham, R.W. Lindsay, and H.J. Read, *J. Electrochem. Soc.*, Vol 12 (No. 1200), 1963, p 112; Vol 4 (No. 401), 1965

Appendix 3.2

Application of Electroless Nickel
in Computer Peripherals*

Introduction

As the demand for bit density of disk drives increases, the drive parameters, such as the flying height of the read/write head above the disks and the head gaplength, decrease to submicrometer levels. For example, in high-density drives, the flying height and the gaplength are about 0.2 to 0.3 μm, and the thickness of the magnetic medium on thin film metallic disks is about 0.05 μm.

With these minute drive parameters, microcontamination seriously affects drive performance, and thus, there is an urgent need to control contamination inside a drive. Both particulate and hydrocarbon contamination could jeopardize the reliability of a drive.

Many of the drive components including the casing are made of cast aluminum alloys, primarily 380 die cast and 356 sand cast alloys. Cast surfaces are notoriously unclean as per the specifications of the drive industry. They can shed particulate contamination and can absorb undesirable hydrocarbons. Epoxy coatings have been used routinely to coat these alloy surfaces. However, the components need to undergo significant precision machining and provide electrical grounding at many areas. Thus, there is an effort to find alternate coatings for cast aluminum alloys.

Electroless nickel is a versatile coating that has many applications in the computer peripherals industry (Ref 1). Experiences with

---

* P.B. Narayan, A.S. Brar, and S. Kaja, presented at *Aluminum Finishing '88*, Conference Proceedings. Reprinted with permission.

electroless nickel as a means of controlling microcontamination on cast aluminum alloy parts used in disk drives is discussed below. Although electroless nickel has much potential to be the coating of choice, much work is necessary, especially in preparing aluminum surfaces for electroless nickel, to realize its full potential.

Adhesion Testing

There is not a suitable test to quantitatively assess adhesion of thin films to substrates. Also, it is necessary to test whether the coated part would shed particulate contamination. The tape test has been chosen to check the suitability of the coating in this present study. In this test, a transparent Scotch® tape is placed on the coating and rubbed manually three times before the tape is pulled out. This is not too demanding of a test for electroless nickel. It is hoped that some work will be done in the industry regarding quantitative tests for adhesion of thin films.

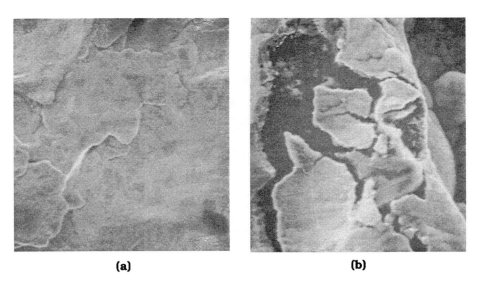

(a)   (b)

Fig. 1   (a) Electroless nickel on a sand cast 356 aluminum alloy showing the rough and uneven growth of electroless nickel (300×). (b) Deficient electroless nickel coating due to surface contamination (1500×).

Sand Cast 356 Alloy Surfaces

Figure 1(a) shows the electroless nickel coated 356 sand cast alloy surface. The surface appears to be rough and has uneven growth of electroless nickel at certain locations. Figure 1(b) represents an area that has deficient electroless nickel coating, probably due to surface contamination on the cast alloy.

Figure 2 shows particles that came out with the tape test, and energy-dispersive X-ray spectroscopy (EDXS) analysis (from Fig. 2a) shows the presence of aluminum, silicon, and iron. These elements are probably the intermetallics in the alloy that were loosely held on the alloy surface.

Figure 2(b) is an example of particles that have aluminum, chlorine, and potassium. They are probably aluminum grains that were loosened during the preplating surface operations (e.g., acid and alkali etching). Figure 2(c) is an example of particles that have calcium and silicon and thus were impurities from the sand mold that stuck to the alloy surface.

A relatively weak signal of nickel on most of these particles indicates that they are at least 5 µm deep. Analysis seems to indicate that there is nothing basically wrong with electroless nickel itself, but the surface preparation made the surface so weak that electroless nickel could not cover it effectively.

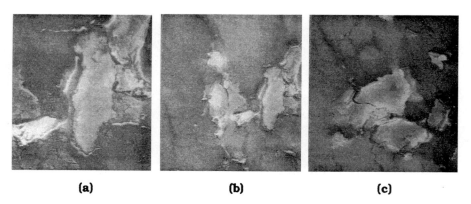

(a)   **(b)**   **(c)**

Fig. 2 Particles that came out of the surface (Fig. 1) during the tape test. The particle in (a) contains aluminum, silicon, and iron (2000×). (b) contains aluminum, chlorine, and potassium (1000×). (c) contains calcium and silicon (1000×).

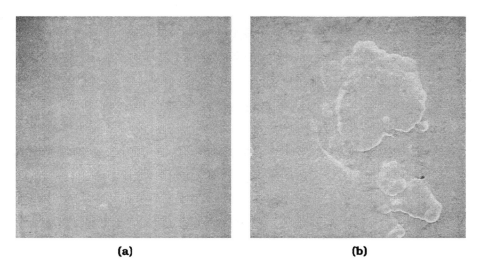

(a) (b)

Fig. 3 Electroless nickel coating on the machined surface of sand cast 356 aluminum alloy (22×). (a) shows the smooth machined surface with the visible machining marks. (b) shows nodular growth at the rough spots of machining (400×).

Machined Surfaces of Sand Cast Alloys

Figure 3(a) shows the machined surface of the sand cast alloy, which has a much smoother surface. However, the machining marks are clearly visible. There appears to be some rough spots near the machining marks that probably gave rise to electroless nickel nodular growth (see Fig. 3b).

Figure 4 shows particles that came out during the tape test. When compared to the sand cast surface, there were much fewer particles that came out with the tape. All the particles contain aluminum, silicon, and iron, indicating that intermetallics seem to be the main cause of particle generation on machined surfaces.

Machining is essentially a very effective cleaning operation that removes the surface contaminants and permits the use of very light chemical etching prior to plating. Thus, the machined surface sheds very few particles. However, intermetallics seem to be separated from the alloy surface during the tape test.

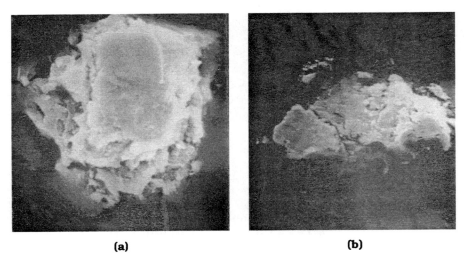

(a)   (b)

Fig. 4 Representative particles from specimen shown in Fig. 3 that came out during the tape test. They contain aluminum, silicon, and iron, the constituents of intermetallics. (a)1500×. (b)1000×.

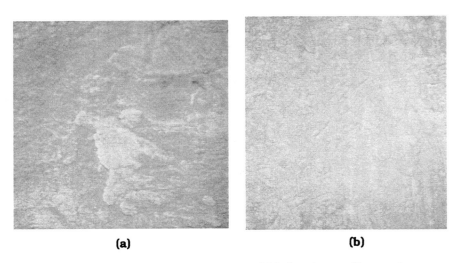

(a)   (b)

Fig. 5 Electroless nickel coated die cast 380 aluminum alloy surface showing the generally smoother surface of die cast parts as compared to a sand cast alloy. (a)170×. (b) 250×.

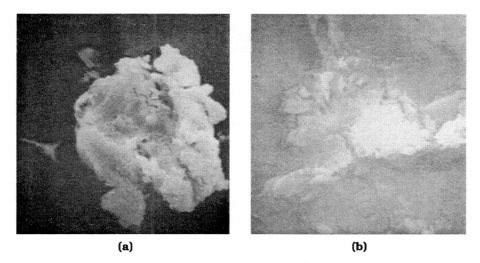

(a)   (b)

Fig. 6 Particles that came out of the surface (Fig. 5). (a) Intermetallics of aluminum, silicon, and iron are visible (2500×). (b) Silicon-base contaminant that stuck on the die cast surface (1000×).

Die Cast Surfaces

Figure 5 shows an electroless nickel coated die cast surface, which is considerably smoother than the sand cast surface. Some of the inhomogeneities on the die surface are replicated by electroless nickel.

Figure 6 shows particles that came out with the tape. The EDXS analysis (Fig. 6a) revealed the presence of aluminum, silicon, and iron, indicating that intermetallics were pulled out during the tape test. The particles appear to have a very smooth surface. Figure 6(b) shows a particle that contains silicon, aluminum, chlorine, potassium, calcium, and iron. It may be a silicon-base particle embedded in the casting that became partly loose during preplating operations.

Summary

There does not appear to be anything seriously wrong with the electroless nickel plating itself. Most of the problems were traced to the preplating/cleaning operations. Sand casting poses the greatest problems as far as adhesion and particle generation are concerned.

Particles embedded from the sand mold appear to be the most common problem of concern.

Die casting produces a smoother and cleaner surface. It sheds fewer particles than the sand cast surface. Machined surfaces shed very few particles. However, intermetallics are a problem here also.

Some effort is needed to develop methods of cleaning sand cast surfaces with minimum chemical etching. For example, bead blasting may mechanically smoothen the surface, thus reducing the need to use heavy chemical etching to prepare the surface for electroless nickel plating. It is also necessary to find ways of hiding or removing intermetallics from the surface so that they cannot be easily pulled from the surface.

Reference

1. S. Kaja, P.B. Narayan, and A.S. Brar, *Products Finishing*, Feb 1987.

Appendix 3.3

Electroless Nickel on Cast Magnesium Alloys in Computer Peripherals*

Use of Magnesium Alloys

In computer peripherals, magnesium alloys are being considered for making critical components because of their excellent weight-to-strength ratio. Because die casting is a cost-effective and convenient technique to manufacture components with critical dimensions, die-cast magnesium alloys are a good choice.

Computer peripherals, with high-information storage capacity, have extremely small dimensional tolerances. For example, the flying height of the read/write head above the disk is 0.25 µm, and the particulate and hydrocarbon contamination levels must be kept very low in the disk drive. It is imperative that the components be kept free from corrosion and particle generation, thus maintaining a clean environment in the memory storage units.

Why EN?

Magnesium is prone to corrosion by oxidation in humid environments at above ambient temperatures. Protective coatings are thus necessary on magnesium to prevent corrosion.

Organic coatings were reviewed, because they provide an excellent barrier to corrosion. However, the part requires post-machining to meet the close dimensional tolerances and mating requirements of

---

* P.B. Narayan and A.S. Brar, presented at *Electroless Nickel '89*, conference proceedings. Reprinted with permission.

the product. The edge of the machined area on a part with an organic coating was found to generate undesirable particulate contamination. The coatings also showed a tendency to flake off at certain locations during machining.

Chromate-type conversion coatings were also reviewed. They showed a tendency to come off easily as their thickness was increased. Especially on die cast surfaces, they exhibited poor penetration resistance against moisture.

Electroless nickel (EN) can plate a thin, uniform, and corrosion-resistant coating with excellent replication of the surface finish of the substrate. Consequently, electroless nickel is the coating of choice to obtain clean and contamination-free surfaces.

Comparison of EN/Al and EN/Mg

Relatively speaking, the industry has a lot of experience in putting a reliable EN coating on aluminum alloys, and there is very little work regarding magnesium. Magnesium does not develop nearly as good a protective oxide scale as aluminum does, thereby making the role of EN more crucial.

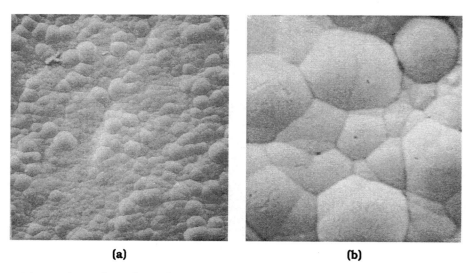

(a)   (b)

Fig. 1 Smooth surface of electroless nickel over a copper underlayer on a cast magnesium alloy surface. (a) 400×. (b) 2000×.

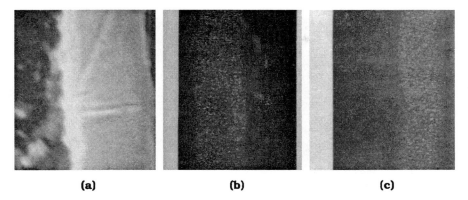

(a)   **(b)**   **(c)**

Fig. 2 (a) Cross section of EN/Cu. Elemental mapping of the cross section for copper (b) and nickel (c) (5000×).

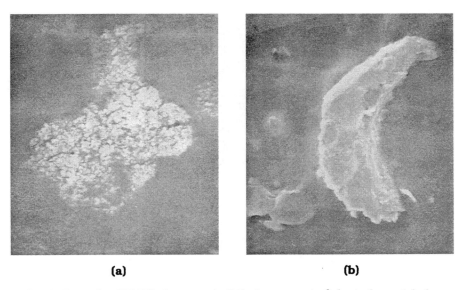

(a)   **(b)**

Fig. 3 Particles (EDXS shows nickel) that came out of electroless nickel layer on magnesium during the tape test, which indicates that the particles were coming from the EN/Mg interface. EN/Cu generated much fewer particles than only electroless nickel during the tape test. (a) 1300×. (b) 600×.

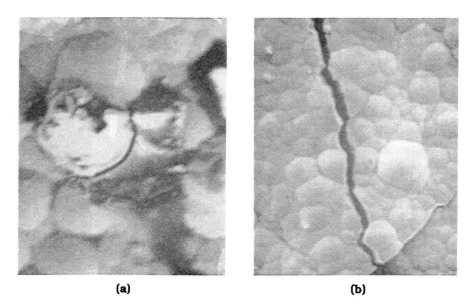

(a)　　　　　　　　　　　　　**(b)**

Fig. 4 Initial stages of corrosion of the electroless nickel layer on magnesium, when the coated part was kept at 65 °C (150 °F) and at 65% relative humidity with 10 ppm of chlorine and SO_2 for 72 h. (a) 2000×. (b) 1000×.

Magnesium is more reactive than aluminum, and its surface preparation by chemical etching was found to be quite challenging. Aluminum alloys contain intermetallic compounds (containing silicon), which pose many problems during EN coating (Ref 1). Magnesium alloys (Mg-Al-Zn) do not have the same problem, but have problems with the segregation of magnesium-aluminum alloy at the grain boundaries.

Testing

Two tests for evaluating the EN coating were used:

- *Tape test:* A transparent sticking tape was put on the surface and rubbed three times manually. The tape was peeled off and was checked to see if any particles came out with the test (Ref 1).

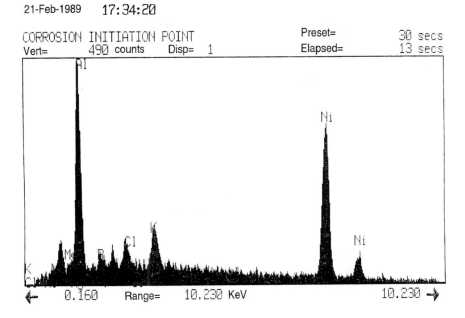

21-Feb-1989   17:34:20

CORROSION INITIATION POINT   Preset=   30 secs
Vert=   490 counts   Disp= 1   Elapsed=   13 secs

0.160   Range=   10.230 KeV   10.230

Fig. 5 EDXS spectrum of the cracked region showing a strong peak of aluminum, indicating corrosion initiation at magnesium-aluminum intermetallics at the grain boundaries.

- *Environmental testing:* The parts were kept in an oven at 65 °C (150 °F) and at 65% relative humidity with several parts per million of chlorine and sulfur dioxide for 72 h and were checked for signs of corrosion. Scanning electron microscopy (SEM) and energy-dispersive X-ray spectroscopy (EDXS) were used to study the morphology and chemical composition of the surfaces before and after environmental exposure.

Experiences with EN

Two types of electroless nickel that were 10 μm thick were studied. One was 10 μm of electroless nickel, and the other was 6 μm of copper underlayer with 4 μm of electroless nickel. It was found that copper underlayer improves the ductility, adhesion, and corrosion resistance of the coating.

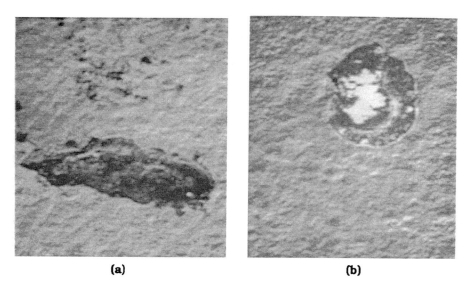

(a)   (b)

Fig. 6 Advanced stage of corrosion showing breaking of the electroless nickel layer, thereby exposing oxidized magnesium. (a) 60×. (b) 100×.

Figure 1 shows the smooth surface of the EN/Cu layer, and Fig. 2 shows the cross section of the coating and the EDXS elemental mappings for copper and nickel.

During the tape test, the number of particles coming out with the tape was much less for magnesium alloys than for aluminum alloys. Even on magnesium alloys, the presence of the copper underlayer drastically reduced the number of particles that came out with the tape. Figure 3 shows some of the particles that came out with the tape. They indicate the presence of only nickel, suggesting that they came from the EN/Mg interface. Of the occasional particle that came out with EN/Cu coating of marginal quality, the particle showed only copper and nickel, suggesting that breakage occurs at the weak copper/magnesium interface.

Figure 4 shows the initial stages of corrosion of electroless nickel in environmental testing. Figure 5 is the EDXS spectrum of the cracked region, showing a strong presence of aluminum. Figure 6 illustrates the advanced stage of corrosion, which indicates that electroless nickel broke exposing the magnesium oxide that formed underneath.

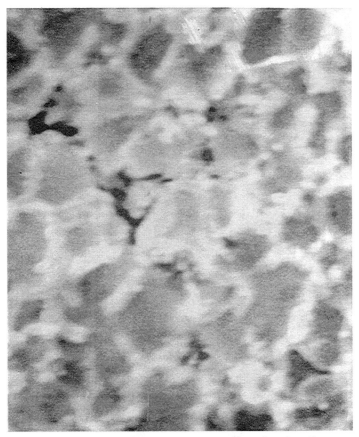

Fig. 7 Microstructure of as-cast magnesium alloy showing the magnesium-aluminum intermetallic compound at the grain boundary. The aluminum-rich area makes electroless nickel brittle and leads to corrosion initiation. 400×.

To understand the corrosion mechanism, let us look at Fig. 7, which is the as-cast surface of the magnesium-aluminum alloy, showing the migration of aluminum to the grain boundaries to form the $Mg_{17}Al_{12}$ compound. The aluminum-rich areas render the electroless nickel brittle and lead to corrosion initiation. Because the aluminum-rich areas are not etched as easily as magnesium, there is differential activation during the surface preparation for electroless nickel, which can affect the local adhesion of the coating.

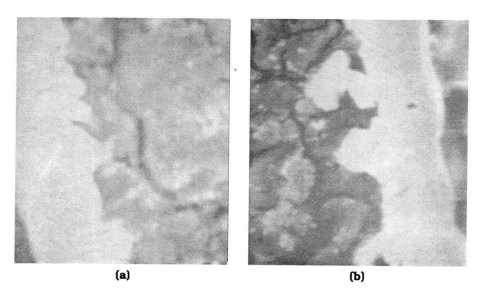

(a)   (b)

Fig. 8 Cross section of EN/Cu over as-cast magnesium alloy, showing how copper fills the surface pores of magnesium, leading better adhesion of electroless nickel to the surface and lower particle generation. (a) 5000×. (b) 2500×.

Also, the differential chemistry of the surface could introduce stress into the deposit. A fine crack in the electroless nickel at the aluminum-rich area provides access for the corroding species, such as chlorine, and moisture to the substrate. Corrosion proceeds at a fast rate, forms a bubble of magnesium corrosion products, and breaks open the electroless nickel at the weak area.

In the case of EN/Cu, this type of corrosion is not observed, presumably because copper does not allow the corroding species to reach the substrate. Also, copper appears to make the EN coating less brittle and more homogeneous.

Figure 8 shows how copper fills the surface pores on magnesium, leading to better adhesion. This action explains why there is very little particle generation when a copper underlayer is present.

Conclusions

During the tape test, particles came out with the tape because of breakage at the EN/Mg interface. Aluminum-rich areas on the as-cast surface make EN more brittle and prone to corrosion. The presence of copper improves adhesion, ductility, homogeneity, and corrosion resistance of electroless nickel. More R & D work is needed to improve the quality and ease of depositing electroless nickel on magnesium alloys.

Reference

1. P.B. Narayan and A.S. Brar, "Application of Electroless Nickel in Computer Peripherals," *Aluminum '88 Finishing*, Gardener Publications, Cincinnati, Ohio, 1988.

Chapter 4

Joining

Manufacturing a product involves joining various materials for a variety of reasons such as electrical continuity, mechanical support, and load sharing. As the components become smaller and more fragile, joining one component to another becomes more challenging. Joining techniques can be divided into three broad categories (Ref 1, 2): metallurgical, adhesive, and mechanical joining.

Metallurgical joining consists of welding, soldering, brazing, and ultrasonic welding. Adhesive joining is achieved with the use of an adhesive (e.g., an acrylic or epoxy resin) that polymerizes or hardens due to stimuli such as ultraviolet (UV) radiation, temperature, absorption of moisture, or chemical reactions. Mechanical joining is achieved with the use of threaded fasteners, rivets, pin fasteners, or special-purpose fasteners such as latches, retaining rings, etc.

4.1 Metallurgical Joining

4.1.1 *Welding*

Welding is a high-temperature process that frequently does not require the use of filler material, in which a bond is made by the melting and solidification of a portion of one, or both, of the joining components. Because considerable melting and solidification occur during the welding process, it is irreversible, and welded samples cannot be reworked. There are several modifications of the welding process, such as resistance, arc, electron beam, gas, flash, and friction welding (Ref 1, 2). The characteristics of resistance welded

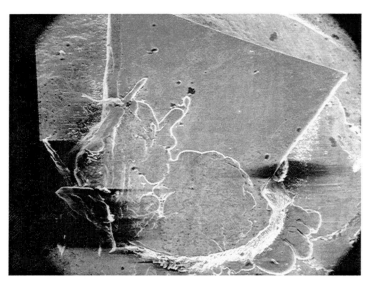

Fig. 4.1 Weld zone showing the violent expulsion of molten metal caused by using excessive welding current. SEM, 54.6×.

stainless steels are discussed in this section, with examples of practical applications and manufacturing case studies.

Resistance welding is a joining process in which two surfaces are joined in one or more spots by the heat generated by resistance to the flow of electric current through the components being joined as they are held together by the electrodes. The contact points are heated by a short-time pulse of low-voltage, high-amperage current, which causes a portion of the component with the lower melting point to melt. The electrodes maintain the compressive force until the molten portion solidifies, thus making the bond. Welding is completed in a fraction of a second.

The major advantages of resistance welding are high speed, suitability for automation, excellent process control, good repeatability, and low unit cost. It also can be used to join similar or dissimilar metals of varying thicknesses.

CASE STUDY 4.1: Control of Welding Current

Problem. Two austenitic (type 300 series) stainless steel components were joined by resistance welding. One lot exhibited premature failure after a few cycles of stress application.

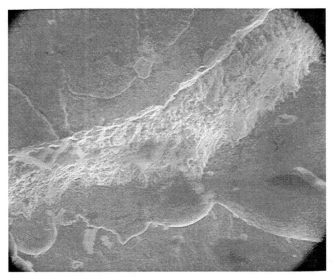

Fig. 4.2 Weld zone showing the chromium carbide precipitation belt around the molten metal. Type 302 stainless steel was sensitized because of overheating caused by too high a welding current. SEM, 273×.

Analysis. Figure 4.1 is a scanning electron microscope (SEM) micrograph of the molten metal in the weld zone, and Fig. 4.2 (at higher magnification) shows the expulsion of molten metal and the chromium carbide precipitation belt surrounding it. Carbide precipitation is visible around the weld zone (Fig. 4.3). The weld failed in the precipitation belt (Fig. 4.4) through cracking (Fig. 4.5) in the precipitation zone (also called heat-affected zone). Figures 4.6 to 4.8 detail the failed area, indicating that failure started at carbide grain boundaries.

Discussion. Visible evidence of violent expulsions of molten metal indicates that the welding current was too high. In production, it is commonplace to use higher welding currents than necessary to produce visible evidence that welding has been completed. In the present case, welding caused undesirable side effects due to the use of too high of a welding current, which led to overheating.

If austenitic stainless steel is kept at approximately 550 °C (1000 °F), carbon in the steel reacts with its chromium content to form chromium carbide precipitates at the grain boundaries. As a result, the areas near the grain boundaries become chromium-depleted and

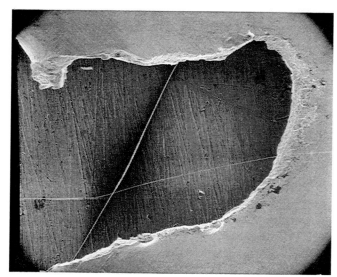

Fig. 4.3 Chromium carbide precipitation around the weld zone. SEM, 78×.

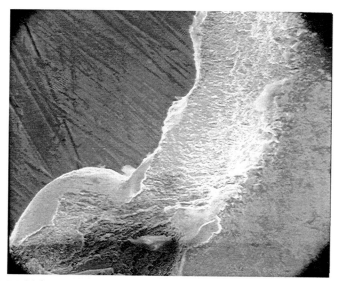

Fig. 4.4 Weld fracture beginning in the carbide precipitation belt. SEM, 312×.

Fig. 4.5 Cracks in the hard, brittle carbide precipitation or heat-affected zone. SEM, 897×.

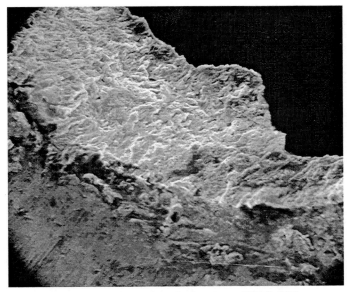

Fig. 4.6 Fractured area showing brittle fracture. SEM, 663×.

Fig. 4.7 Beginning of fracture at carbide grains. SEM, 624×.

Fig. 4.8 Initiation of cracking at the brittle carbide grains. SEM, 624×.

are susceptible to corrosion. Also, because carbides are hard and thus brittle, premature cracking can initiate in this area. This form of degradation of stainless steel is called sensitization. If the temperature is kept above 850 °C (1600 °F), the carbides become dissolved in the matrix. Because sensitization, which is controlled by the solid-state diffusion of chromium in the matrix, is time-dependent, minimizing the amount of time the steel is kept in the sensitizing temperature zone is imperative.

Recommendation. To prevent this type of failure, the welding current should be lowered, and statistical process control procedures should be implemented to control the welding process. Lower welding current decreases the heat generated and thus the size of the heat-affected zone. Welding parameters should be optimized by monitoring the strength of the joint. The design of the joint should also be improved by making the two flanges bigger (see Fig. 4.9) so that the flange will not act as a pivot point for the spring and the stress will be distributed more evenly over the entire area.

CASE STUDY 4.2: Underheating of the Weld and Subsequent Reworking

Problem. A resistance weld joint of two stainless steel components experienced a high rate of premature failure.

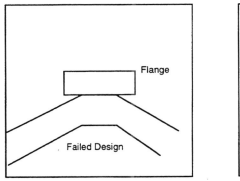

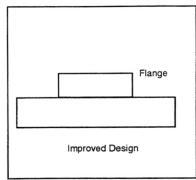

Failed Design   **Improved Design**

Fig. 4.9 Schematic of the flange design that failed and the improved design, which distributes stress more evenly.

Analysis. Visual examination of the failed joints indicated that at least 80% of the joints were reworked. There appeared to be very little melting of the metal. In about 40% of the failed samples, the weld was made in an off-centered location. Five samples were mounted in resin and polished to obtain a cross section of the weld sections. Figure 4.10 shows cross sections where there was no weld penetration, whereas Fig. 4.11 depicts a good weld with sufficient weld penetration. In Fig. 4.12, the weld sections contain voids and porosity, which are indicative of poor welding practice. Figure 4.13 illustrates off-centering of the weld joint, which results from improper surface preparation during reworking.

Discussion. Welds with less than 20% penetration are known to completely fail due to normal variations in current, time, and electrode force. Insufficient penetration during welding can be caused by unoptimized welding parameters such as insufficient welding current, insufficient welding time, improper heat balance, and improper surface conditions.

Voids and porosity in the weld sections are caused by improper welding force, improper rate of welding current increase, insufficient welding time, and expulsion of metal due to current surges. During

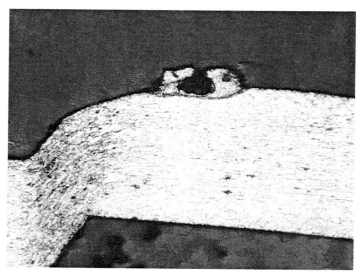

Fig. 4.10 Unacceptable weld cross section showing very little weld penetration, with minimal weld strength. 74.8x.

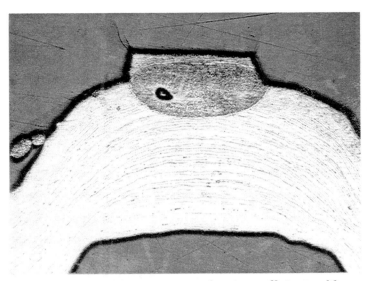

Fig. 4.11 Acceptable weld cross section showing sufficient weld penetration. 74.8×.

Fig. 4.12 Unacceptable weld cross section showing voids and porosity caused by use of inappropriate welding parameters. 74.8×.

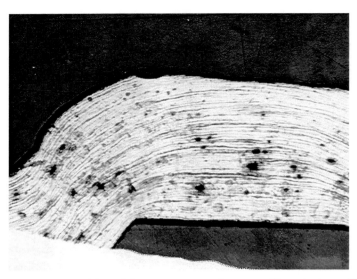

Fig. 4.13 Unacceptable weld cross section showing off-centering of the weld joint caused by harsh mechanically abrasive cleaning during reworking. 74.8x.

reworking, parts often need to be cleaned by harsh mechanical abrasion, which could lead to uneven and slanted surfaces and thus to the off-centering of the welding electrodes and the weld joints.

Recommendation. Welding parameters should be optimized by monitoring weld strength. Reworking appears to cause many problems and, consequently, should be controlled more closely. Reworked samples should be separated from other specimens in production and should be inspected carefully at each stage of welding. Reworking should be minimized, or preferably, eliminated.

CASE STUDY 4.3: Evaluation of Welding Parameters

Problem. Two austenitic stainless steel pieces were joined by resistance welding. The quality of the subsequent bond was evaluated.

Analysis. To evaluate the weld joint, three different parameters were reviewed: (1) weld strength, (2) microstructure and weld penetration, and (3) repeatability or reproducibility.

Weld Strength: The pull strength in tension shear varied from 20 to 54 kgf (45 to 118 lbf). During pull testing, almost all of the samples failed by shearing of the weld metal. The weld strength was much lower than that of the parent metal.

Microstructure: Weld cross sections were prepared by mounting the samples in resin and polishing. Figure 4.14 shows cross sections of poor welds, indicating that insufficient melting and shear failures occurred throughout the weld nugget. The weld section in Fig. 4.15 exhibits good penetration of the weld on both sides and shear failure around the weld.

Reproducibility: Figures 4.14 and 4.15 demonstrate that there is great variability in the quality of welded samples. The average pull strength of about 15 samples was 39 kgf (86 lbf), with a standard deviation of 9.5 kgf (21 lbf), 25% of the average.

Discussion. Most samples failed consistently in the weld zone, rather than in the parent metal, during pull testing, indicating that the welds were unsatisfactory. Also, the pull strength of the weld joint was considerably lower than the parent metal, which is also indicative of poor weld quality. Weld cross sections showed insufficient melting and inadequate penetration at the welding interface. The wide variations in pull strength of the welded joints indicated a lack of process control.

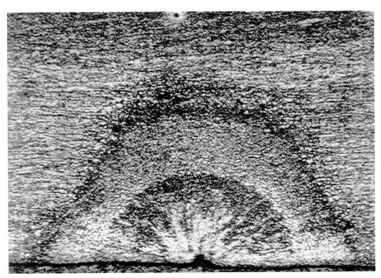

Fig. 4.14 Cross section of unacceptable weld showing insufficient melting and shear failure through the weld nugget. 117×.

Fig. 4.15 Cross section of acceptable weld showing deeper penetration
and shear failure around the weld. 150×.

Recommendation. It was recommended that the welding parameters be optimized based on the pull strength. After optimization,
statistical process control should be implemented.

CASE STUDY 4.4: Weld Qualification

Problem. To qualify a vendor who made resistance weld joints of
two austenitic stainless steel components, a study of the process
variables and resulting welds was undertaken. The vendor selected
the welding parameters and provided sample welds made by both ac
and dc welding. The relative merits of the two techniques were
compared.

Analysis. The joints were evaluated from the standpoint of weld
strength, microstructure and weld penetration, and reproducibility.
These variables are discussed individually below.

Weld Strength: Weld strength was measured by a tensile testing
machine. The ac weld samples exhibited higher strength, an average
of 5.2 kgf (11.5 lbf) than the dc weld samples, an average strength of
4.4 kgf (9.7 lbf).

Fig. 4.16 Cross section of an ac weld exhibiting adequate melting and penetration. 96×.

Microstructure: Cross sections of the weld joints were prepared by mounting the samples in resin and polishing. Figures 4.16 and 4.17 show the cross sections of ac weld and dc weld joints, respectively.

Reproducibility: Standard deviations of the weld strengths for the ac weld and the dc weld joints were 11 and 12%, respectively.

Discussion. It was determined that all weld strengths were satisfactory, even though the ac weld strength was 20% higher. Both cross sections showed that satisfactory melting and penetration had occurred (more than the 20% penetration threshold for acceptable welds). Reproducibility was good for both groups. The outer surfaces of the joint, where the welding electrodes came in contact with the components, were smooth and free of overstressing, indicating that the compressive load during welding was correctly suited to the application. However, the dc weld samples exhibited slight discoloration and overheating, indicating that the dc welding voltage should be reduced.

Recommendation. Both groups of welds appeared visually satisfactory; but the ac welds were better. If dc welding is to be used, welding parameters need to be optimized.

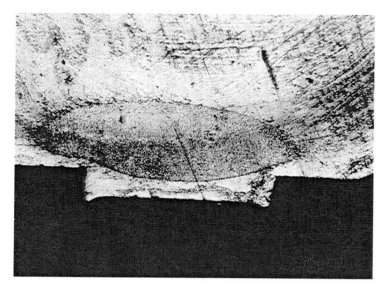

Fig. 4.17 Cross section of a dc weld exhibiting adequate melting and penetration. 96×.

CASE STUDY 4.5: Weld Failure Caused by Edge Roughness

Problem. Two resistance welded joints of type 304 full-hard stainless steel exhibited premature field failures. In addition to the failed components, two joints of the same service age were studied for comparison. One of the components was manufactured by defining the edges with a chemical etching process.

Analysis. The failed components were observed in a low-power (100×) optical microscope. The size of the weld nuggets was adequate, and there appeared to be sufficient melting during welding. There was nothing apparent that indicated a potential problem with the weld.

The weld areas and edges were examined using SEM analysis. Figure 4.18 shows the rough, porous edges with overstressed regions, and Fig. 4.19 illustrates the fractured surface and the area in which the resistance weld broke. An example of the used joints from the same lot that experienced failure is shown in Fig. 4.20. Note that the edges prepared by chemical etching are rough, porous, and weak.

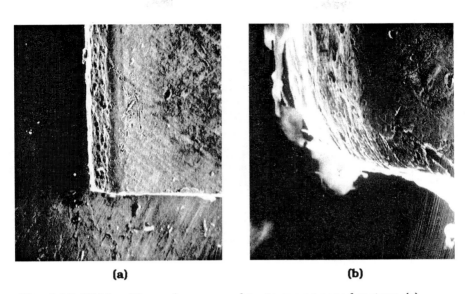

(a)	**(b)**

Fig. 4.18 Welds with rough, porous edges in overstressed regions. (a) SEM, 100×. (b) SEM, 1000×.

Fig. 4.19 Weld fracture surface. SEM, 200×.

Discussion. The chemically etched edges were rough and porous, and the notched areas at the edges acted as stress concentrators. During stress application, cracks could initiate at these points and propagate through the main body of the weld. The welding process itself was satisfactory. The sample produced was full-hard type 304 stainless steel, which is not very ductile. Consequently, the crack that started at the edge propagated through the main body and weld area, thus causing fracture.

A decision was made to machine-punch the steel specimens instead of etching them chemically. Figure 4.21 shows the edges of the machine-punched samples from Vendor A, with smooth and stress-free edges, whereas Fig. 4.22 is a machine-punched sample from Vendor B that has rough and highly stressed edges. Factors that contributed to these edge defects include the use of improper material for the punch die, excessive die wear, misalignment of the die, or improper load application.

Recommendation. Because chemical etching causes porous, rough edges that could lead to fracture, it should not be used.

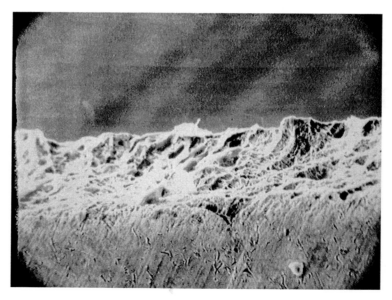

Fig. 4.20 Rough, porous edges of samples prepared by chemical etching. SEM, 500×.

Fig. 4.21 Machine-punched sample with smooth and stress-free edges. SEM, 1000×.

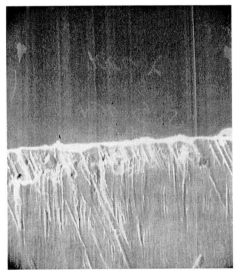

Fig. 4.22 Machine-punched sample with rough and highly stressed edges. SEM, 1000×.

Machine punching, when properly performed, is a superior alternative to chemical etching from the standpoint of edge integrity and cost control.

4.1.2 *Soldering and Brazing*

Soldering and brazing are primarily used to join two metals with the help of a filler metal that has a lower melting point than either of the materials to be joined. In general, the filler typically used in soldering has a melting point below 300 °C (approximately 600 °F), whereas the filler typically used in brazing has a melting point above 425 °C (approximately 800 °F). As a result, soldering relies primarily on wetting of the metals by the solder for joint strength, whereas brazing relies on both wetting and solid-state diffusion.

Of the metallurgical and adhesive joining techniques, only soldering (and occasionally brazing) is a truly reversible process. By applying sufficient heat, it is possible to open the joint and separate the two parts after soldering. Consequently, soldering has the significant advantage of flexibility in replacing, repairing, and reworking components without destroying the main part. Because brazing involves significant solid-state diffusion, opening of the joint is much more difficult and is often not recommended.

Soldering is commonly used when numerous joints must be made quickly at low cost. Thus, it is well suited to automation. The use of soldering for completing simultaneous connections of multiple terminals on a printed circuit (PC) board is an excellent example of typical production applications for this process.

In the microelectronics industry, aluminum, copper, silver, and gold are common metals to be joined, and soldering is the choice for joining. Brazing is used primarily to join steels, a subject that is beyond the scope of this discussion.

During soldering, the liquidus temperature of the filler (or solder) should be 50 to 100 °C (90 to 180 °F) below the solidus temperature of the substrate. Eutectic alloys are often used as solder material, because they have lower melting points than either of the alloy elements.

Important considerations in the choice of a solder material include a low melting point, high electrical conductivity, high mechanical strength of the joint, materials compatibility from the standpoint of galvanic corrosion, and, of course, reasonable cost. For example, in tin-lead alloys, higher lead content increases mechanical strength and decreases cost, but the melting temperature increases, leading to increased soldering time and expense. Higher heat dissipation at high soldering temperatures could damage delicate components.

The popular solder materials in the order of increasing melting point are alloys based on bismuth, tin, indium, lead, zinc, and cadmium. Cadmium is no longer considered a candidate material because of concerns about its toxicity. Because of concerns that lead accumulation in the brain causes mental retardation, an effort is underway to minimize, and possibly eliminate, the use of lead in all industries, including the joining industry.

Critical concerns in the soldering process are (1) surface preparation, including fluxing, (2) amount of solder, (3) heating of the solder, and (4) the cleaning procedure to remove flux and other unwanted chemicals from the solder joint area.

The main objective of surface preparation is to remove oils, greases, and oxide scale by chemical etching (e.g., solder flux) or mechanical abrasion, because their presence makes the metal surfaces chemically inert and reduces wetting by the solder. Fluxes contain acid, salts, or amines that are capable of removing the oxide layer. For a satisfactory joint to be completed, a direct atomic contact should exist between the solder and the metal. Consequently, any film-like contamination and the oxide layer (or sulfide or similar tarnish film) should be removed from the metal surface.

Use of insufficient solder does not provide adequate surface coverage and consequently does not provide sufficient joint strength. Excess solder adds unwanted weight to the component and acts as a heat sink to reduce the soldering temperature, thus causing underheating. The rate of wetting increases with temperature. Underheating thus decreases the strength of the joint. Overheating decreases throughput, increases oxidation problems, and can cause thermal damage to the component. Thus, the amount of solder and the solder temperature should be optimized for the process.

After soldering is completed, the flux should be removed from the surface of the joint because of its corrosive action. Because the use of aqueous cleaning is expected to replace chlorofluorocarbon (CFC) cleaning, improvements in the soldering process include the use of water-soluble fluxes and elimination of flux usage altogether.

Traditionally, copper and its alloys have been primary candidates for soldering. Currently, aluminum and its alloys are finding increasing use because of their formability, high conductivity, and excellent passivity characteristics. As a result, copper-aluminum soldering junctions are becoming commonplace.

There are two important factors that should be considered when soldering copper and aluminum—wetting of both with the same solder and selection of the appropriate solder (Ref 1). Aluminum is passivated by ambient exposure due to the formation of an oxide film, which prevents good wetting by the solder. The aluminum surface should be mechanically or chemically cleaned just prior to soldering. Wherever possible, pretinned aluminum leads should be used, because a strong bond already exists at their aluminum-solder interface, which is the weakest link in the solder joint.

What is an appropriate solder for use with aluminum? Eutectic tin-lead and 60-40 tin-lead are commonly used for copper, but they do not wet aluminum very well. Also tin-lead is very electropositive in the galvanic series compared to aluminum, and thus their junction is prone to galvanic corrosion, particularly in humid environments. Consequently, zinc-containing solders (e.g., 95-5 zinc-aluminum and 91-9 tin-zinc eutectic) are normally used with aluminum.

CASE STUDY 4.6: Use of Inadequate Solder

Problem. Copper flex leads were being pretinned with the 63-37 eutectic tin-lead alloy solder (63% tin and 37% lead). One particular lot exhibited poor joint strength after soldering to another copper conductor.

Hypothesis. Because the soldering process was under strict control, the in-coming pretinned flex leads were suspected as the origin of the problem.

Analysis. Acceptable and unacceptable lots of flex leads were examined with an optical microscope. The acceptable lot contained more solder material, which was also brighter and shinier. The samples were studied using SEM and energy-dispersive X-ray spectroscopy (EDXS). Figure 4.23 is a SEM micrograph of an acceptable lot of solder, which exhibits the typical eutectic structure. Alloys of eutectic composition have a sharp and well-defined freezing point, and at the freezing temperature the two phases (corresponding to the eutectic composition) are precipitated out. Because there is insufficient time for any appreciable diffusion, the two phases are very finely distributed. Figure 4.24 is a SEM micrograph of the solder from the unacceptable lot. There are clearly two phases present—one that is lead-rich and another phase that is tin-rich (close to eutectic

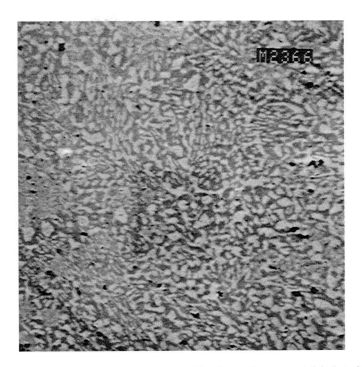

Fig. 4.23 SEM micrograph of Pb-Sn solder from the acceptable lot showing the characteristic eutectic structure.

composition). Because the surface microstructure is not that of an eutectic alloy, the surface composition is clearly off-eutectic.

The edges of the solder area, where the solder was thin (see Fig. 4.25), of a good pad were examined.This area exhibited an off-eutectic surface structure, similar to Fig. 4.24. The thick central portion had a bright eutectic structure (see Fig. 4.26).

Discussion. During the solder reflow operation, tin diffuses into the copper, thus making the solder layers adjacent to the copper tin-poor, or lead-rich. If the solder is not thick enough (as was the case with the unacceptable lot), the tin-poor layers can extend to the surface. When this off-eutectic alloy freezes, it forms two phases and consequently has lower brightness. The off-eutectic alloy melts at a

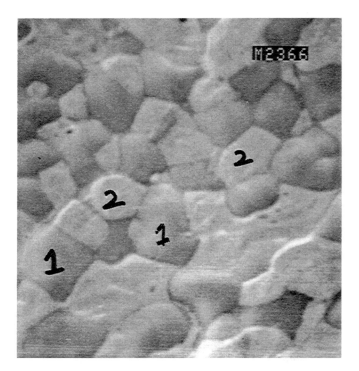

Fig. 4.24 SEM micrograph of the surface of Pb-Sn solder from the unacceptable lot with an off-eutectic structure with lead-rich (1) and tin-rich (2) phases.

temperature that is higher than the eutectic temperature. Consequently, the off-eutectic solder does not melt during soldering.

Note that soldering parameters were set assuming a eutectic composition and temperature. If the solder does not melt completely, joint strength will naturally be lower. In the acceptable lot, the solder layers adjacent to the copper became depleted in tin, but the solder was so thick that the surface composition remained unchanged at the eutectic level. The eutectic solder melted properly during soldering, forming a strong bond. The fact that off-eutectic composition was observed at the edges of the good solder pad confirms the observation that, if the solder layer is thin, the surface develops an off-eutectic composition.

Solution. It was recommended that the solder layer have sufficient thickness so as not to deplete the surface layers of tin.

Fig. 4.25 Off-eutectic Pb-Sn structure visible at the edges of an acceptable solder pad.

CASE STUDY 4.7: Compositional Difference vs Brightness

Problem. A white haze was noticed on the lead-tin solder area present around a through-hole on a printed circuit board.

Hypothesis. A chemical reaction (e.g., corrosion) or overheating was suspected of causing the discoloration.

Analysis. Figure 4.27 is a SEM micrograph showing the haze around the perimeter of the hole. The brightness contrast with the surrounding area was much more striking in an optical microscope. Figure 4.28(a) is a high-magnification micrograph of the affected area, and Fig. 4.28(b) illustrates the normal area. In Fig. 4.28(a), the white band is lead-rich and the dark band is tin-rich, thus indicating the presence of an off-eutectic two-phase structure. Figures 4.29(a) and (b) provide the EDXS spectra of the areas shown in Fig. 4.28(a)

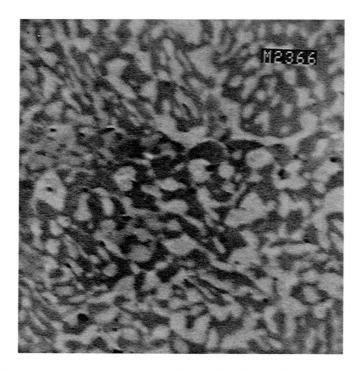

Fig. 4.26 Eutectic Pb-Sn structure observed in the thick central region of an acceptable solder pad. The edges of this area exhibited an off-eutectic structure.

and (b), respectively, demonstrating that the affected area is basically lead-rich.

Discussion. There was no evidence of corrosion or overheating in the affected area. Because that area is lead-rich and has an off-eutectic two-phase structure, optically it appears to be hazy. The holes were drilled slightly undersize and were plated-through with lead-tin eutectic solder. During insertion of the pins through the holes, excess solder was pushed out and collected around the hole. This may have caused the formation of the hazy area.

Solution. Because there was no corrosion and only slight compositional difference from the rest of the area, the blemish was found to be innocuous and did not affect the functionality of the printed circuit board in this application.

Fig. 4.27 White haze around a through-hole on a printed circuit board.

CASE STUDY 4.8: Insufficient Stirring of the Solder Pot

Problem. It was necessary to join a copper conductor to an aluminum strip with 91 to 99% tin-zinc eutectic alloy solder. Copper was supplied pretinned with the tin-zinc solder. One particular lot exhibited poor joint strength.

Analysis. The pretinned copper appeared to have a few patches on the solder surface during visual examination. The samples were studied using SEM and EDXS. Figure 4.30 is a SEM micrograph showing two distinct phases, one zinc-rich and the other having the eutectic composition. Figures 4.31 and 4.32 illustrate the EDXS

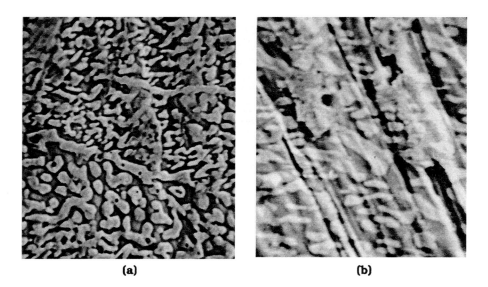

(a)   (b)

Fig. 4.28 Surface structure of (a) the hazy area with the off-eutectic Pb-Sn
structure and (b) the normal surrounding area with the eutectic Pb-Sn
structure.

spectra of the zinc-rich island and the surrounding eutectic area,
respectively.

Discussion. The dipping operation was studied, and it was found
that the pot was not being stirred at regular intervals. Because zinc
is lower in density than tin, it tends to accumulate on the surface,
thus enriching the surface layers in zinc. When the copper leads are
dipped in the solder, zinc-rich, off-eutectic alloy is plated on the
copper. This alloy has a higher melting point than the eutectic 91-9
tin-zinc alloy and thus does not melt properly and uniformly during
soldering, thereby rendering the solder joint weak.

Solution. Frequent stirring of the tin-zinc solder pot produces
uniform composition throughout the pot and facilitates plating of the
eutectic alloy on the copper leads.

CASE STUDY 4.9: Improper Cleaning

Problem. As in Case Study 4.8, aluminum strip was supplied
from the vendor with 91-9 tin-zinc alloy on the end leads. One

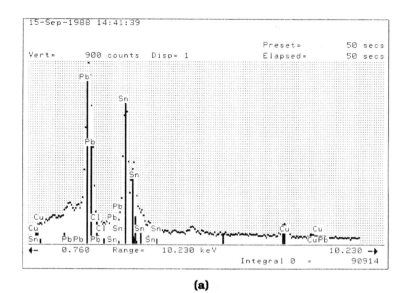

(a)

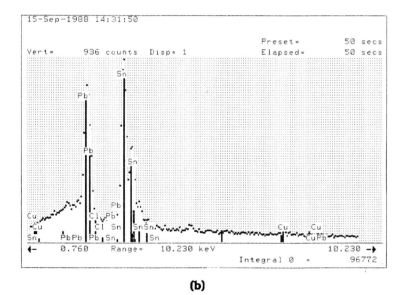

(b)

Fig. 4.29 EDXS spectra of (a) hazy area and (b) normal area, revealing higher amounts of lead in the hazy area.

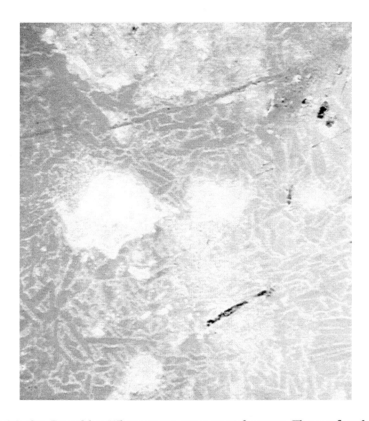

Fig. 4.30 Sn-Zn solder. White spots are zinc-rich areas. The surface has an off-eutectic structure. SEM, 800×.

particular lot exhibited dark spots on the solder. It was necessary to determine whether the lot was acceptable as supplied.

Analysis. Figure 4.33 is the EDXS spectrum obtained from the spotted region, which indicates primarily the presence of zinc and chlorine.

Discussion. The solder had an off-eutectic composition, and thus, excess zinc was present on the surface, as in Case Study 4.8. The presence of chlorine was due to the flux or the cleaning procedure; the chlorine reacted readily with the excess zinc. The cleaning procedure was inadequate and ineffective. A small amount of copper was also contained in the dark spot (Fig. 4.33), which could have been caused by cross-contamination from the cleaning procedure.

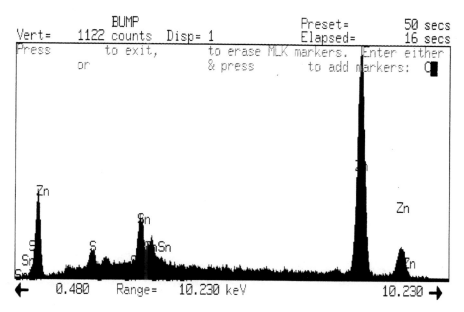

Fig. 4.31 EDXS spectrum of a typical white patch that exhibits higher zinc content than the surrounding area.

Chlorine is very corrosive and could adversely affect long-term reliability. Consequently, the presence of chlorine cannot be tolerated.

Solution. It was recommended that the cleaning procedure used by the vendor be improved to eliminate the presence of chlorine. The lot containing the dark spots was rejected.

CASE STUDY 4.10: Incomplete Removal of Oxide

Problem. It was necessary to join copper to aluminum with 91-9 tin-zinc alloy. The aluminum leads came from a coil wound with an anodized aluminum wire. The anodized oxide layer provided an insulation barrier so that the coil did not short-circuit. The oxide layer was ground off with a sand paper drill before soldering. One particular lot exhibited poor joint strength after soldering.

Analysis. Several of the unacceptable joints were peeled apart, and the exposed surfaces were studied by SEM and EDXS. The aluminum side of the joint surface was covered with a ridged anodized layer (see Fig. 4.34). The cracked layers shown in the micrograph are remanents of the anodized layer that adhered to the

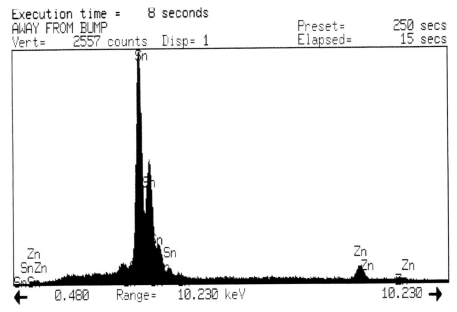

Execution time =   8 seconds
AWAY FROM BUMP   Preset=   250 secs
Vert=   2557 counts   Disp= 1   Elapsed=   15 secs

Fig. 4.32 EDXS spectrum of normal 91Sn-9Zn solder from an area adjacent to that shown in Fig. 4.31.

aluminum. No copper was observed on the aluminum side of the joint. The copper side of the joint showed evidence of the solder and the transfer of the aluminum oxide layer (see Fig. 4.35a and b for SEM micrographs from two different samples). Figure 4.36 provides another view of the copper side of the joint, and Fig. 4.37 illustrates elemental mapping for aluminum for that area.

Discussion. The joint fractured at the aluminum-alumina interface, and there was no wetting by the solder on the aluminum side of the joint, which indicates that the alumina layer was not properly, or completely, removed. There was no interdiffusion between the solder and the aluminum, thus rendering the joint weak.

Solution. The anodized aluminum oxide layer should be completely ground off before soldering.

CASE STUDY 4.11: Optimization of Aluminum Oxide Removal

Problem. As part of the solution in Case Study 4.10, it was necessary to determine how to optimize the removal of the aluminum oxide barrier layer from the aluminum conductors.

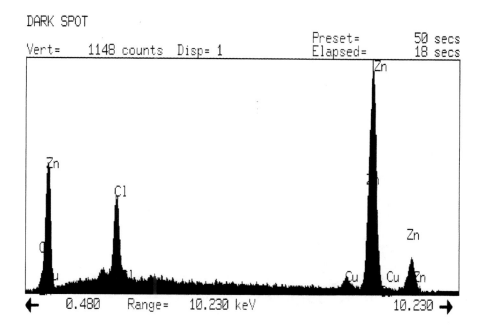

Fig. 4.33 EDXS spectrum of typical dark spot on end leads of aluminum strip showing the presence of zinc and chlorine. The zinc-rich area corroded.

Discussion. If a coarse grit size is used on the sandpaper drill, the oxide removal rates are higher, but the surface becomes rough and some islands of oxide remain on the surface. If a very fine grit size is used, the oxide removal rate is slow, but the surface will be relatively smooth without islands of oxide. How can a compromise be struck between the two considerations?

Solution. A coarse grit size should be used first to remove the majority of oxide scale. Then fine grit size can be used to smooth the surface by removing any remaining oxide. It is necessary to change the grit paper frequently so that it does not become clogged with debris.

CASE STUDY 4.12: Passivation of Aluminum

Problem. Similar to Case Study 4.10, it was necessary to solder copper and anodized aluminum with 91-9 tin-zinc solder. The anodized layer was ground off with a sandpaper drill. Even though the

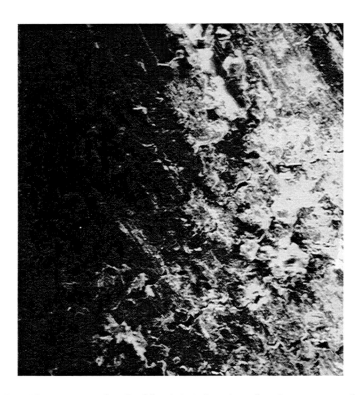

Fig. 4.34 Aluminum side of solder joint showing aluminum covered with ridges of cracked anodized layer.

oxide layer was removed properly, one particular lot exhibited poor joint strength.

Analysis. Several of the joints from the unacceptable lot were peeled apart, and the exposed surfaces were studied by SEM and EDXS. The copper side of the joint had a considerable amount of solder, whereas the aluminum side showed no transfer of solder.

Discussion. The fact that solder was not visible on the aluminum side of the joint indicates that there was no wetting of solder on the aluminum and the surface was not adequately prepared. However, it was apparent that the anodized oxide layer was removed properly from the aluminum surface. After observing the actual manufacturing process, it was discovered that there was frequently a time lag of several hours between the removal of the oxide layer and the actual

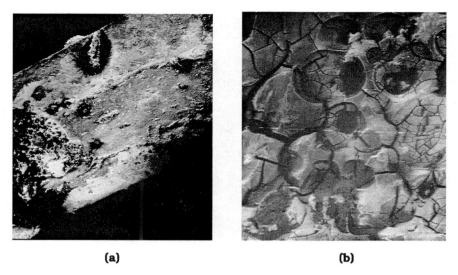

(a)   (b)

Fig. 4.35 (a) Copper side of the joint showing the transfer of anodized aluminum onto copper. (b) Anodized aluminum on copper, indicating that shearing occurred at the aluminum-alumina interface.

soldering operation. Several hours of atmospheric exposure led to the formation of a passive layer of aluminum oxide, which prevented wetting of the solder.

Solution. The manufacturing process was modified slightly so that the anodized oxide layer was removed only a few minutes before soldering, thereby minimizing the passive aluminum oxide layer. Figure 4.38 depicts the freshly ground aluminum strip, showing aluminum metallic grains, and Fig. 4.39 illustrates how well the Sn-Zn solder wets the freshly ground aluminum. After several days of exposure to atmospheric air, the ground aluminum begins to form inert and insulating aluminum oxide (Fig. 4.40), which leads to charging and image distortion in SEM analysis. The white streaks are the passive aluminum oxide layers formed due to atmospheric exposure. Figure 4.41 illustrates improper wetting and the consequent beading of solder on the inert aluminum surface.

CASE STUDY 4.13: Handling Damage

See Appendix 4.1 for an in-depth discussion of an attempt to reduce handling damage in soldered wire.

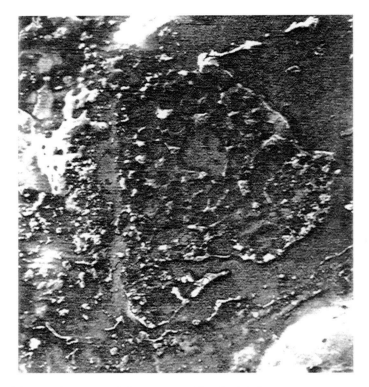

Fig. 4.36 Another view of the copper side of a joint.

4.1.3 Ultrasonic Bonding

During ultrasonic bonding, an axial compressive load is applied to the area where the joint is to be made, and ultrasonic energy is transmitted through the horizontal motion (in a plane parallel to the weld area) of the tip. Ultrasonic vibrations cause elastic hysteresis, localized slip, and plastic deformation, and the oxide layers on the components are fractured by surface shear forces created by the high-frequency vibrations. Also, frictional heat is generated at localized contact points on the two surfaces. All of these factors contribute to an effective bonding of the two surfaces.

The energy generated by ultrasound is proportional to the thickness and hardness of the material, as in the empirical equation:

Fig. 4.37 Aluminum mapping of area shown in Fig. 4.36 confirming the transfer of anodized aluminum onto copper.

$E = 63\,H^{n}t^{n}$, where $n = 3/2$

E is the acoustic energy in joules, H is the Vickers hardness number, and t is the material thickness in inches. This relationship is valid if t is less than 0.8 mm (0.03 in.).

No thermal damage occurs during ultrasonic bonding. Bonding is accomplished without introducing external heat or passing electrical current through the components to be joined. Heating, caused by this technique, is highly localized and does not cause thermal damage to areas adjacent to the bonding area. Bonding time ranges from a fraction of a second to 2 s.

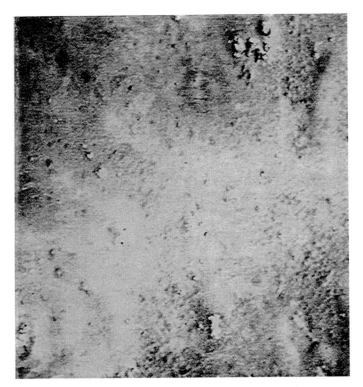

Fig. 4.38 Freshly sanded aluminum strip showing metallic grains.

No surface preparation is necessary. This technique does not require the use of flux or any other surface preparation. The ultrasonic vibrations remove the oxide layers, hydrocarbon layers, and other contaminants.

No postcleaning operations are needed. Because ultrasonic bonding does not use flux and does not leave any crevice for entrapment of foreign materials in the bonded area, there is no need to clean the area after joining. Because there is very little microstructural change during bonding, the corrosion resistance of the bonded materials in and around the weld area do not change.

Ultrasonic bonding offers wide versatility in production applications. With sufficient power, it is possible to weld metals covered with thin lacquers, enamels, anodic coatings, oils, and other foreign

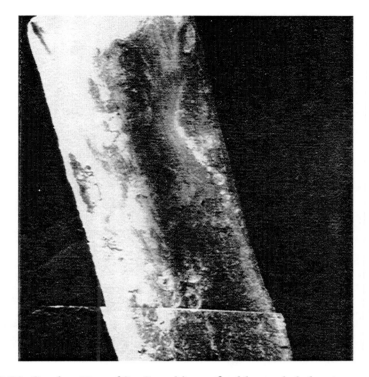

Fig. 4.39 Good wetting of Sn-Zn solder on freshly sanded aluminum.

materials. However, it is difficult to bond metals that are laminated to plastic films.

CASE STUDY 4.14: Use of Too Much Weld Load

Problem. An anodized aluminum component was being soldered to a copper strip using 91-9 Sn-Zn eutectic alloy. In an effort to replace soldering with ultrasonic bonding, several samples were prepared for evaluation.

Analysis. Both the soldered and bonded samples were found to have similar resistance (about 0.5 Ω) and peel strength. For microstructural analysis, the bonded samples were peeled apart and studied by SEM and EDXS. Figure 4.42 shows aluminum wire spread over copper. The load appears to be too large, thus rendering

Fig. 4.40 Passive aluminum oxide formed on aluminum due to extended
atmospheric exposure. Streaking and image distortion are due to charging
caused by the insulating oxide.

the aluminum wire weak. The aluminum surface that made the bond
is shown in Fig. 4.43. There is no transfer of copper onto the
aluminum, suggesting that there was no melting of copper. During
peeling, fracture evidently occurred in the aluminum. Figure 4.44
shows the copper surface with small islands, approximately 50 μm
(1970 μin.) in size, of aluminum sticking to the copper.

Discussion. The ultrasonic bond between the untinned copper
and the anodized aluminum exhibited low electrical resistance and
good peel strength. The bonding process appeared to be quite effective in removing the anodized oxide layer from the aluminum surface.
The weld load appeared to be too high, with opportunity for optimizing bonding parameters.

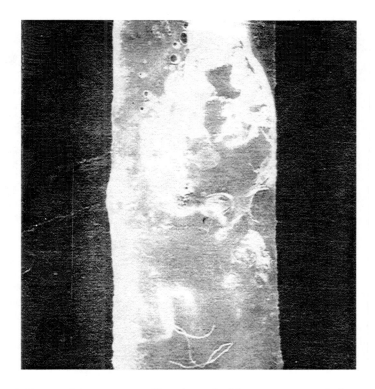

Fig. 4.41 Incomplete wetting and beading of solder on passivated aluminum.

Recommendation. Bonding is a satisfactory replacement for soldering. However, the weld load was too high. It should be decreased, while simultaneously increasing the bonding temperature (by increasing either the ultrasonic frequency or the bonding time).

CASE STUDY 4.15: Evaluation of Diffusion Joints

Problem. An anodized aluminum component was bonded to a copper strip using ultrasonic bonding. The bonding parameters were optimized based on peel strength. An evaluation of any adverse microstructural changes was undertaken.

Analysis. The bond was pulled apart in the shear mode, and the sheared surfaces were studied by SEM and EDXS. Figure 4.45 is a SEM micrograph of the copper side of the bond showing aluminum

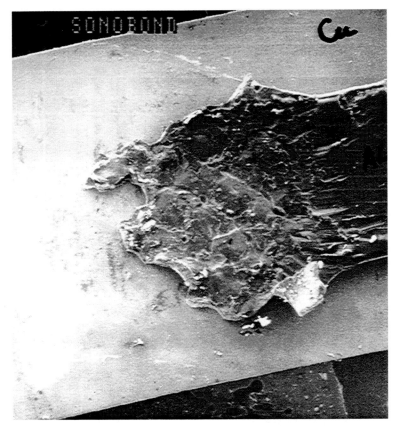

Fig. 4.42 Aluminum wire deformed and spread over copper during ultrasonic bonding. The weld load was too high.

islands. Examination of the aluminum surface indicated that aluminum grains were pulled from the surface during the shearing process. There was no transfer of copper onto the aluminum.

There was a definite increase in the width of the aluminum wire, which was caused by a combination of plastic flow in the aluminum and by application of compressive load. Clearly, there was some softening of aluminum. Because aluminum has a lower melting point (660 °C, or 1220 °F) than copper (1083 °C, or 1981 °F), plastic

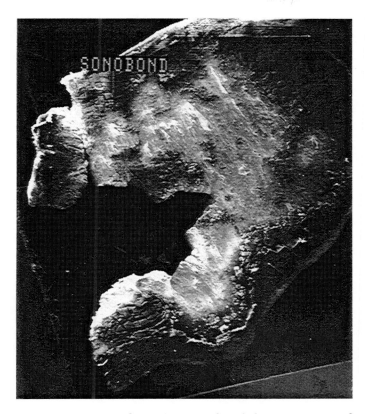

Fig. 4.43 Aluminum side of the ultrasonic bond showing no transfer of copper because of the higher melting point of copper.

deformation and softening in aluminum can be expected to occur more readily at lower temperatures than in copper.

On the copper side, whenever aluminum was detected, a small but significant amount of sulfur was also found. This sulfur was probably the result of the sulfuric acid anodizing of aluminum prior to bonding.

The bonded strips were mounted vertically in a plastic mold and then ground and polished through the middle. Examination of the cross sections indicated that the bond between the aluminum and copper was strong and free of defects.

Fig. 4.44 Aluminum islands on the copper side of the joint showing that aluminum diffused and bonded during the bonding procedure.

Discussion. During bonding, the aluminum became sufficiently hot at localized points to flow plastically, and it diffused into the copper at the compression points of bonding. Because of the higher melting point of copper, there was insignificant diffusion or melting of copper. During shear failure, the breakage started at the interface between the aluminum and the copper-aluminum diffuse layer and progressed through the aluminum grains by shear. A significant amount of sulfur remained, presumably from the aluminum anodizing operation. This sulfur should be removed by an effective cleaning operation (such as ultrasonic cleaning in water). There was no damage to the material in the surrounding area, as expected.

Fig. 4.45 Copper side of a strong ultrasonic joint showing islands of aluminum. Aluminum diffuses much faster than copper due to frictional heat generated during bonding.

Recommendation. Ultrasonic bonding leads to significant diffusion of aluminum into the copper and provides excellent bonding and low contact resistance. However, aluminum should be cleaned thoroughly before bonding to remove all traces of sulfur from the anodizing operation.

CASE STUDY 4.16: Evaluation of Bonding Procedure

Problem. Bonding of a copper wire to a gold-plated pad produced occasional failure.

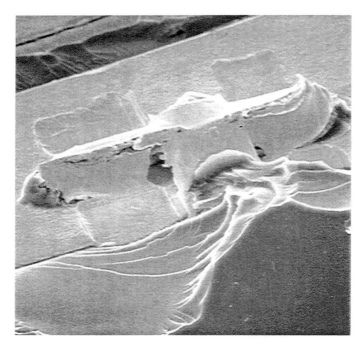

Fig. 4.46 Failed junction of copper on gold pad during ultrasonic bonding showing evidence of multiple attempts at bonding with the probable presence of film-like contamination at the interface.

Analysis. Figure 4.46 is a SEM micrograph of the failed junction area where evidence of multiple attempts at bonding is visible. Figure 4.47 of the neighboring junction does not reveal an open circuit, but it is evident that the wire became quite weak and was ready to fracture.

Discussion. The gold pads were probably covered with hydrocarbon contamination, making the ultrasonic bond difficult to complete. In the process of making multiple attempts to complete the bond, the wire became very weak, particularly at the heel of the ultrasonic tip. One junction fractured due to wire movement, and the other junction was ready to break as well.

Recommendation. If a bond cannot be completed under normal conditions, hydrocarbon contamination on the bonding surface

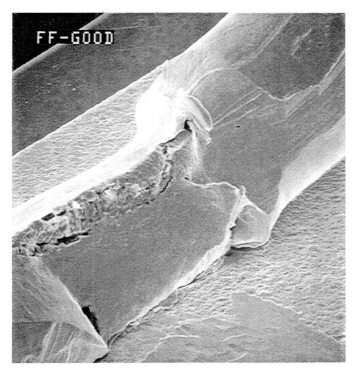

Fig. 4.47 Neighboring junction showing severely weakened copper wire ready to fracture.

should be investigated. The load during bonding must be reduced so that the wire does not become severely deformed.

4.2 Adhesive Bonding

Adhesive bonding involves the use of an organic resin, such as an epoxy or cyanoacrylate, that is cured at the site of the junction between two mating parts. Two-part resins, in which the resin and hardener are mixed just prior to application, can be cured at ambient temperature, because the hardener facilitates the cross-linking of the resin. One-part resins required stimuli such as heat or UV radiation for cross-linking (or curing) to be achieved. With the development of new and convenient adhesive systems, there has been a dramatic increase in the use of adhesive bonding in all industries. For example, in the automotive industry, 20 to 30% of all joints involve adhesive bonding.

Advantages of adhesive bonding are numerous. It uses ambient or only moderately high temperatures (about 100 °C, or 212 °F), and as such there is no thermal damage to the joining parts. Adhesive bonding can be used to bond metals, ceramics, and plastics in any combination. It does not involve the use of flux, which is quite corrosive if allowed to remain in the joint area. Special postcleaning operations are not required.

Adhesive bonding does not cause any dimensional change in the product because the adhesive layer can be as thin as a few tens of nanometers. It is a relatively simple and inexpensive operation. Some of the precautions needed to ensure successful adhesive bonding are surface preparation, proper adhesive curing, and control of potential outgassing. These are discussed below.

4.2.1 *Surface Preparation*

For optimum adhesion, the joining surfaces must be rough (so as to increase the area of contact between the adhesive and the surface) and clean (devoid of any contamination even at the molecular level). If at all possible, the surface should be roughened, e.g., by mechanical abrasion. Mechanical abrasion also removes strongly adherent mill scale and metallic oxide scale. A metal surface without an oxide layer facilitates a strong joint, because metal/adhesive bonding is much stronger than metal/oxide bonding.

Particulate contamination is relatively easy to remove by mechanical scrubbing, ultrasonic washing, high-pressure jet cleaning, etc. Hydrocarbon or film-like contamination is more insidious, because it is much more difficult to detect and eliminate.

There are many sources of film-like contamination. Industrial atmospheres contain hydrocarbons such as greases, pump oils, and machining oils. Mold release agents are a common processing contamination on die cast parts. Human handling introduces contamination due to spittle marks, fingerprints, body oils, and facial make-up ingredients. Leaching from plastic solvent containers or use of impure solvents is another source of contamination. Water is widely used in manufacturing, and thus, water retention or adsorption and water spots (deposition of mineral salts dissolved in water) are sources of contamination.

CASE STUDY 4.17: Contamination Removal With Gaseous Plasma

Problem. In the manufacture of read/write heads for disk drives, the ceramic head slider is bonded to the stainless steel plate with an adhesive. Occasionally, it was found that the bond strength was very low.

Analysis. Several bonds were broken in the shear mode and observed by SEM. The bond consistently broke at the adhesive-plate interface.

Figure 4.48 shows the auger electron spectroscopy (AES) spectrum of a typical plate surface. Because stainless steel derives its corrosion resistance from the presence of a thin, approximately 10 nm (0.394 µin.), adherent and passive chromium oxide layer, surface enrichment of chromium is anticipated, with the presence of large peaks of chromium and oxygen and small peaks of iron and nickel. However, Fig. 4.48 shows a significant peak of carbon, indicating the presence of a thin layer of hydrocarbons. This layer remained on the surface despite cleaning with organic solvents.

Discussion. To evaluate the bonding procedure, the curing temperature and curing time for the adhesive were reviewed. They were found to be within specification. The adhesive was thus ruled out as a possible cause of failure. Because the bond broke at the plate-adhesive interface, the presence of contamination on the plate was suspected, which was confirmed by AES analysis (Fig. 4.48).

Subsequently, the plates were cleaned for about 10 min in a gaseous plasma. Figure 4.49 shows the AES spectrum of the surface after plasma cleaning, indicating a drastic reduction in carbon content, or removal of the hydrocarbon layer. When the adhesive bond was made between the plasma-cleaned plate and the slider, bond strength was found to be quite high.

Recommendation. In future applications, the metal surface should be cleaned in gaseous plasma immediately before bonding to ensure good bond strength.

4.2.2 *Adhesive Curing*

The adhesive curing characteristics (such as curing temperature, time of exposure, etc.) should be determined using thermal analysis techniques (see Appendix 1.1). They should be confirmed by peel

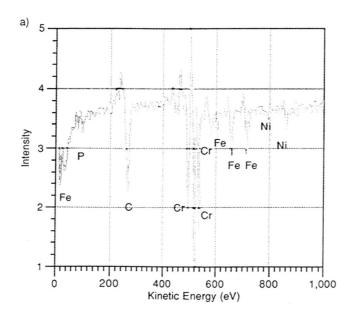

Fig. 4.48 AES spectrum of type 302 stainless steel surface with film-like hydrocarbon contamination.

strength measurements. The oven temperature profile, quantity of batch load, presence of different materials that could affect curing, and time required for oven-temperature equilibration are some of the experimental conditions that need to be monitored to ensure optimum resin-curing parameters.

CASE STUDY 4.18: Undercured Copper Coil Assembly

Problem. The actuator assembly of a particular disk drive consisted of copper windings held together by a heat-cured epoxy. In one lot of such assemblies, a track-seek error was discovered during final product checkout procedures. This occurred when the assembly was programmed to go to a particular track location, but it went to an adjoining track instead.

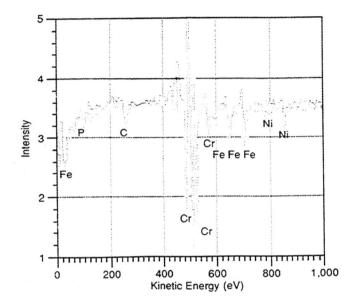

Fig. 4.49 AES spectrum of plasma-cleaned surface showing removal of hydrocarbon contamination, which led to an increase in bond strength.

Analysis. The actuator coil moves very rapidly under the applied magnetic field, and bonding between the bobbin, windings, and the outer ring should be extremely strong. It was suspected that the bobbin was sagging, thus changing the location of the head. Thermogravimetric analysis of the epoxy bond indicated that the bond was losing weight when it became heated, suggesting that it had not been fully cured during normal processing.

Discussion. Incomplete epoxy curing led to a weak bond and a resultant sagging bobbin. Using a thermal analysis system, cure temperature and time were optimized for the epoxy in the presence of the copper wire winding. When the new specifications were introduced into the manufacturing process, the problem of erratic coil performance disappeared.

In addition, metallographic analysis of coil cross sections revealed gaps in the winding (see Fig. 4.50). Although the track-seek error was associated with the epoxy bond of the actuator, the coil winding procedure was modified to compact the windings (see Fig. 4.51) and improve the performance of the actuator.

Recommendation. New specifications for the resin-curing temperature and time should be introduced. The coil winding process was modified to compact the windings.

4.2.3 *Potential Outgassing*

Optimum adhesive-curing temperature and time should be determined not only to maximize strength, but also to minimize or eliminate outgassing. The outgassing component can condense on a compatible surface and create particulate contamination.

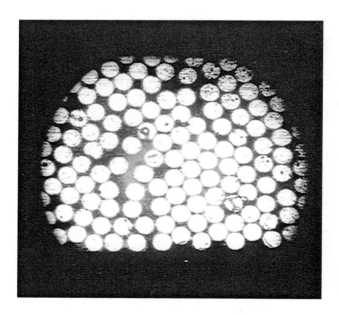

Fig. 4.50 Metallographic cross section of a copper wire coil showing gaps between wires and inefficient winding.

4.3 Mechanical Joining

There are many types of mechanical fasteners. Among the nondestructive types are threaded fasteners, pin fasteners, and special-purpose fasteners such as retaining rings, latches, slotted springs, and studs. Destructive-type fasteners include rivets and blind fasteners. Typical failures of these joints are covered in detail in *Metals Handbook, Failure Analysis and Prevention,* Volume 11, Ninth Edition. For more information, see the chapter on Mechanical Failures in this book.

One undesirable aspect of mechanical joining, generation of particulate microcontamination, is discussed in the following section in terms of threaded fasteners such as screws. Their potential for generating particulate contamination is so high that a hallmark of good design in the computer peripherals and microelectronics industry is use of the smallest number of screws. Generally, austenitic

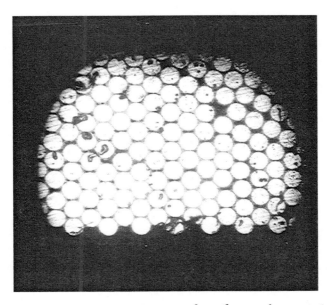

Fig. 4.51 Metallographic cross section of a coil wound more tightly, leading to improved bobbin performance.

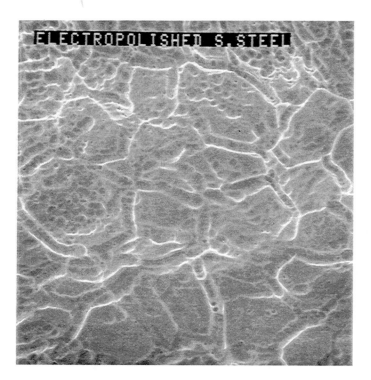

Fig. 4.52 Electropolished stainless steel with a clean, contamination-free surface.

stainless steel screws are preferred because of their good corrosion resistance.

4.3.1 *Use of Electropolishing to Deburr Screws*

The process of cutting threads on screws generates slivers, and sometimes they remain loosely attached to the screws. Electropolishing is an efficient technique to remove slivers and loose contamination from screws (Ref 3).

Electropolishing is basically the reverse of electroplating. In electropolishing, surface layers are removed from the screws because they are used as anodes in a low-voltage plating cell that uses an acidic or alkaline electrolyte. During electrolysis, the products of anodic dissolution react with the electrolyte to form a polarized film

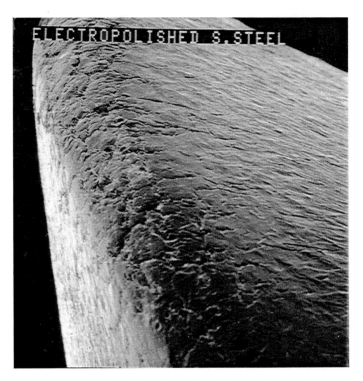

Fig. 4.53 Smooth, sliver-free edge of the threads of an electropolished screw.

that conforms to the broad contours of the surface. At microroughness peaks, the film is thinner. Consequently, the electrical resistance is lower, and the current is higher at those localized points. As a result, the current is very high surrounding the slivers, and they are dissolved much faster than the remainder of the screw. During this process, the surface becomes smooth at the micro-level.

The size of the burr that can be removed depends on the tolerance of the part. Sometimes, it may be necessary to mechanically polish the surface first and then smoothen it by electrodeburring. The latter procedure requires a current density of 5380 to 21,500 A/m^2 (500 to 2000 A/ft^2), whereas electropolishing requires 540 to 5400 A/m^2 (50 to 500 A/ft^2).

Because stressed areas are dissolved on the surface, an electropolished surface has the same galvanic potential at various points,

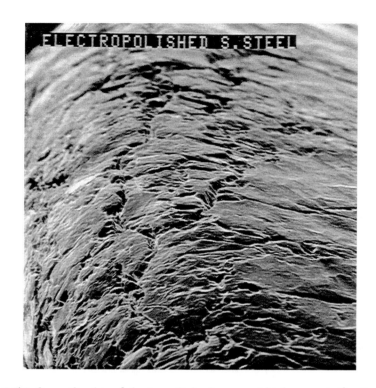

Fig. 4.54 Smooth edge of electropolished screw at higher magnification.

and the tendency to form galvanic cells is minimized. A thin, passive oxide layer is formed on the stainless steel surface due to electro-polishing. Figure 4.52 shows a typical clean surface, with mildly etched grain boundaries, obtained on stainless steel by electropolishing. Figures 4.53 and 4.54 are low- and high-magnification SEM micrographs of the smooth and sliver-free edges of the threads of an electropolished screw.

4.3.2 *Use of Electroless Nickel Coating to Prevent Gouging*

When stainless steel screws are driven into and taken out of an aluminum alloy part, there is a significant amount of gouging and material removal, which generates particulate contamination. When a thin layer of electroless nickel is deposited on the screws, gouging is eliminated due to the lubricity of the electroless nickel (Ref 4).

References

1. H.H. Manko, *Solders and Soldering*, 2nd ed., McGraw-Hill, 1979
2. H.H. Manko, *Soldering Handbook for Printed Circuits and Surface Mounting*, Van Nostrand Reinhold, 1986
3. L.J. Durney, Ed., *Electroplating Engineering Handbook*, 4th ed., Van Nostrand Reinhold, 1984
4. S. Kaja, P.B. Narayan, and A.S. Brar, Electroless Nickel for Computer Disk Drives, *Products Finishing*, Feb 1988, p 46

Appendix 4.1

Fracture Analysis of Soldered Wire by Plasma Ashing and SEM*

The increasing demand for higher quality and reliability in the computer peripherals industry requires that the components function without premature materials failure. Scanning electron microscopy has been found to be an indispensable tool for studies leading to improved materials and processes. Because of the variety of materials used, SEM specimen preparation required continuing innovation.

Low-temperature oxygen plasma-ashing was found to be a convenient technique to remove organic resins and other hydrocarbons (Ref 1). A gaseous plasma, consisting of neutral, energized and ionized atoms, is created by passing oxygen through a high-voltage electric field. The plasma readily reacts with the hydrocarbon, forming gaseous carbonaceous by-products. The addition of argon to oxygen provides sputter etching and a more effective resin removal. The temperature of the specimen rarely goes above 50 °C (120 °F), and consequently it is called low-temperature gas plasma. The plasma does not alter surfaces or bulk properties of metals and ceramics.

A particular failure analysis case is described below, in which plasma ashing and SEM were successfully used to solve the fracture

---

* P.B. Narayan, *Proc. 11th Int. Cong. Electron Microscopy*, Vol 4, *Materials Sciences*, San Francisco Press, 1990, p 638-639. Reprinted with permission.

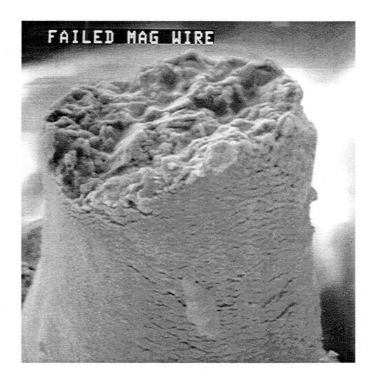

Fig. 1 Fractured wire indicating necking of the wire before fracture.

failure of a soldered copper wire, thereby increasing the reliability of the whole unit.

The lack of electrical continuity in the circuitry of a few disk drives led to close visual examination that indicated that the wire carrying the electrical signal from the read/write head to the drive electronics showed premature fracture. The 50 μm diameter wire was made of electrolytic copper and was soldered to an anchor pad with eutectic lead/tin solder. The solder joint was covered with an organic resin cured by UV radiation. The wire fracture was located underneath the transparent resin. A more puzzling finding was that some of the failed units showed intermittent contact in electrical testing and no apparent fracture in visual examination.

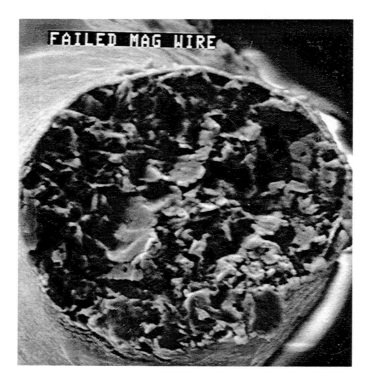

Fig. 2 Fractured rough surface showing ductile failure.

Any mechanical means of removing the resin cover, such as scraping, cross-sectioning, or lapping was found to alter the fracture morphology. Low-temperature oxygen plasma was found to remove the resin nondestructively so as to expose the fractured surfaces for SEM examination.

Figure 1 shows necking due to plastic deformation of the broken wire, whereas Fig. 2 shows the rough and dull fractured surface indicating a ductile fracture. Therefore, wire embrittlement, either at the vendor site or during soldering, was ruled out as the failure cause.

Figure 3 shows microcracking through the solder material and evidence of mechanical stressing of the wire at its junction with the

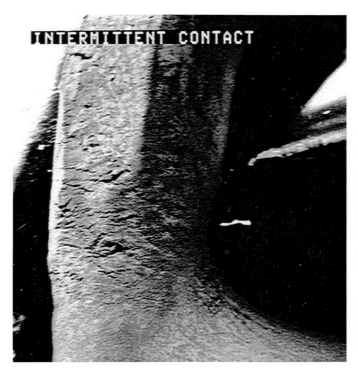

Fig. 3 Wire joint at the anchor pads showing stress.

anchor pads. The stressing was apparently caused by wire handling after soldering. Figure 4 is the fractured surface of the wire with intermittent contact. The presence of resin is clearly visible on the fractured surface, which indicates that the fracture occurred before the resin was cured and the broken wire was held together by the resin. Any movement of the wire during handling could make or break the contact, thereby creating intermittent contact.

Elimination of wire movement after soldering and modification of the resin application process (so as to reduce handling damage) solved the failure problem. The mechanical properties of the wire, both as received from the vendor and after soldered, were found to be satisfactory.

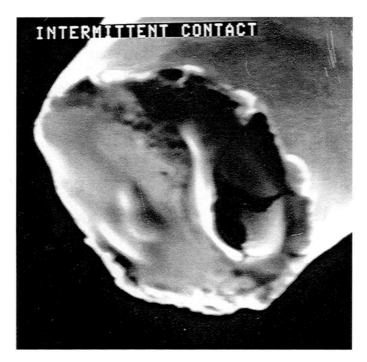

Fig. 4 Fractured surface of the wire with resin.

Reference

1. P.B. Narayan and A.S. Brar, *Connection Technology*, Oct 1988, p 19-21

Chapter 5

Mechanical Fracture

To better understand the concept of mechanical failure, consider an engineering stress versus tensile strain diagram for a material (see Fig. 5.1). In the elastic region, strain increases linearly with stress, the slope being the Young's modulus, E. The stress-strain behavior in this region is dependent on the material, but not on the processing route. Beyond the elastic region, plastic deformation sets in until the

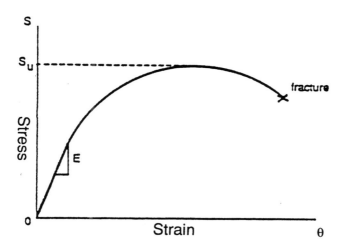

Fig. 5.1 Typical engineering stress vs tensile strain plot for a ductile material such as a metal.

ultimate tensile strength is reached, at which point localized deformation (or necking) begins. Fracture occurs at fracture stress.

Figure 5.2 shows typical stress-strain diagrams for several materials. As the hardness of steel increases, the onset of plastic deformation is delayed. At the same time, the amount of plastic deformation decreases, and fracture occurs at much lower strains. Cast iron and timber exhibit the typical behavior of brittle materials (e.g., ceramics) with negligible plastic deformation.

Most metals exhibit some plastic deformation, the extent of which depends on processing history, surface roughness, the presence of embrittling species such as hydrogen and chlorine, the amount of cold work, etc. Microstructurally, fractured surfaces look rough and dull after plastic deformation, indicating intragranular failure, whereas brittle fractured surfaces appear smooth, shiny, and faceted, which indicates intergranular cleavage.

As some of the case studies below indicate, a ductile material such as steel can fail like a brittle ceramic if the steel is charged with hydrogen, thus causing hydrogen embrittlement.

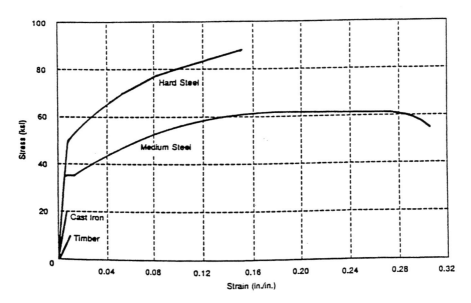

Fig. 5.2 Stress vs strain diagram for ductile materials such as steel and brittle materials such as cast iron and timber.

CASE STUDY 5.1: Effect of Surface Roughness on the Strength of Copper Traces

Problem. A flex circuit exhibited premature failure.

Analysis. Optical microscopy examination of the flex circuit revealed that one of the copper traces in it possibly fractured. An innovative technique using low-temperature gas plasma was developed to remove the plastic layers without any mechanical damage to the fractured surfaces. The SEM examination revealed that the trace had indeed fractured and that even the undisturbed surface had cracks. Surface profilometry revealed that the traces were unusually rough.

Solution. Copper traces were chemically etched very aggressively when the flex circuit was manufactured. Because of the rough and uneven surface, cracks developed during flexing, and the trace fractured. It was also found that the flex circuit was attached with a metal clip to a die cast component with several protrusions on its surface. These protrusions acted as stress concentrators, thereby facilitating premature fracture of the trace.

Recommendation. The traces should be chemically etched much less aggressively. Either the surface protrusions on the cast part should be removed, or the path of the flex circuit should be redesigned so as to avoid that component. See Appendix 5.1 for further details.

CASE STUDY 5.2: Bending Through a Sharp Radius

Problem. During shipping of disk drives, the drive belt is held in place by a retainer clip. During product development, the failure rate of this clip was found to be unusually high. The clip is made from high-carbon steel that is hardened, formed into shape, and zinc coated for corrosion protection.

Analysis. Under scanning electron microscopy (SEM) examination, the zinc-plated surface appeared normal, with no signs of over-stressing apparent. It was observed that fracture occurred at the point of sharp bending. Figures 5.3(a) and (b) show the fractured surface and its zinc elemental mapping by energy-dispersive X-ray spectroscopy (EDXS), respectively.

Solution. Zinc elemental mapping of the fractured surface reveals that zinc penetrated into the steel, which indicates that the crack existed before the zinc plating operation. As a result, zinc plating was ruled out as a possible source of the problem. The high-carbon steel

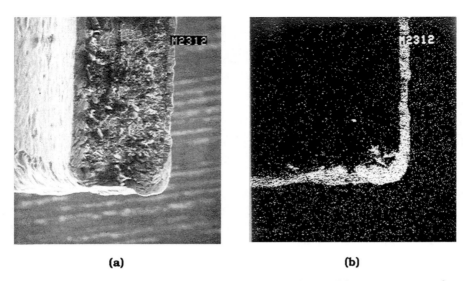

(a)   (b)

Fig. 5.3 (a) Fractured surface of zinc-plated steel part. (b) Zinc mapping of
the same area showing that the crack existed prior to zinc plating.

exhibited high hardness, and when it was bent through a sharp
radius, the stress exceeded the fracture strength.

Recommendation. Forming stresses should be reduced by bending the clip through a larger radius. If the same radius is required,
use of a steel with a lower carbon content is recommended, which
will produce lower hardness, but higher fracture strain.

CASE STUDY 5.3: Hydrogen Embrittlement During Zinc Plating

Problem. Zinc-plated music wire was used as a torsion spring for
the read/write head in some drive designs. The springs from one
pre-production lot broke easily during assembly.

Analysis. The fractured surface appeared to be smooth, faceted,
and shiny. An SEM examination clearly established that it was a
brittle fracture.

Solution. Hydrogen embrittlement was suspected. However, the
processing specifications included baking the part at 200 °C (390 °F)
for 1 hour after zinc plating to remove hydrogen. To obtain optimum
hydrogen removal, the part should be baked within 1 hour of zinc

plating. It was found that baking was performed sometimes after allowing parts to sit overnight or over a weekend. After such a long delay, baking will not remove all of the hydrogen. High-carbon steels with hardnesses of 34 HRC or higher are especially prone to hydrogen embrittlement, because they absorb the atomic hydrogen generated during zinc plating very quickly.

Recommendation. The part should be baked within 1 hour after plating. The plating schedule was modified accordingly.

CASE STUDY 5.4: Effect of Decarburization on Hardness

Problem. A critical drive component made of steel worked loose and was rattling because one of its fastener lock washers had become deformed.

Analysis. Microhardness measurements were taken on a cross-sectional sample of the washer. The center had a hardness of 48 HRC and the surface a hardness of 30 HRC. When the cross section was etched, a light-colored band was found near the surface.

Solution. The light-colored band was evidence of decarburization. When the carbon content of the steel decreased, its hardness decreased. Consequently, the surface layers could not withstand the applied load and failed. The steel washer was heat treated to obtain the required hardness, but if the heat treating environment contained oxygen, the latter would react with the carbon in the steel, leading to decarburization.

Recommendation. The heat treating environment was changed to a reducing atmosphere, to prevent decarburization.

CASE STUDY 5.5: Improper Materials Selection and Machining

Problem. One lot of connector pins experienced a high premature failure rate. The pins were made of copper and plated with nickel and gold.

Analysis. Representative samples from acceptable and unacceptable lots were encased in a plastic mold and polished to measure their hardness, which was found to be 38 HRC. Figures 5.4 and 5.5 show the microstructures of the acceptable and unacceptable samples, respectively. Although the unacceptable sample exhibited a slightly larger grain size, both had essentially the same microstructure of beryllium copper. Chemical analysis confirmed that both lots

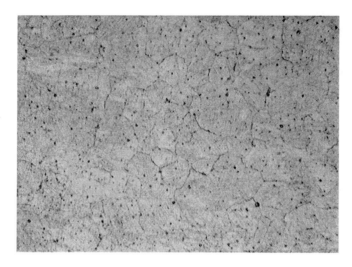

Fig. 5.4 Microstructure of an acceptable connector pin, confirming that the material is beryllium copper. 156×.

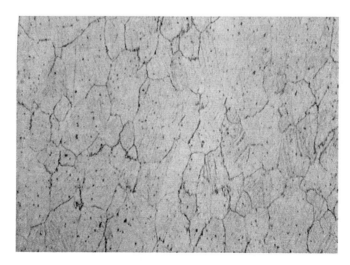

Fig. 5.5 Microstructure of an unacceptable connector pin. It is essentially similar to that of an acceptable pin. 156×.

(a)   (b)

Fig. 5.6 SEM micrographs of pins showing machine indentation marks and cracks emanating from them.

were made of beryllium copper. Figures 5.6(a) and (b) are SEM micrographs of two samples from the unacceptable lot.

Machine indentation marks near the slots and cracks emanating from the slot are clearly visible in these micrographs.

Solution. Similar microstructure and hardness values indicated that there was no difference between the quality of the material from the acceptable and unacceptable lots. The machine indentation marks on the unacceptable samples indicated that, during the threading operation, the pin was gripped near the slot instead of on the solid section. Because the material has a hardness of 38 HRC, cracks developed near the slot due to gripping at an improper place. The cracks could easily propagate during pin insertion, causing premature failure.

Recommendation. The gripping procedure was changed so that the part was held on the solid section of the pin during threading. Because there was no need to use such a high hardness material, beryllium copper was replaced with brass as the pin material. Brass has much higher ductility and does not allow easy crack initiation and propagation.

CASE STUDY 5.6: Improper Materials Selection

Problem. One lot of screws made of 1020 plain-carbon steel exhibited premature failure.

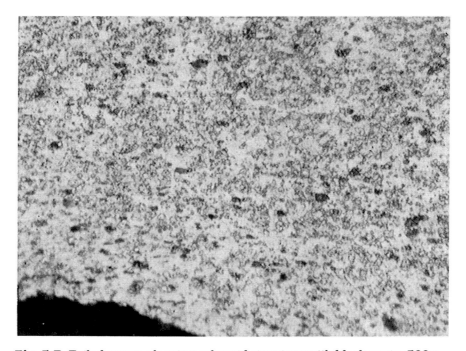

Fig. 5.7 Failed screw, showing spheroid structure with black spots. 500×.

Analysis. A failed screw was mounted in a plastic mold, polished, and etched to study its microstructure. It had a spheroid structure with black spots (see Fig. 5.7). Figure 5.8 is an SEM micrograph showing the black spots, which revealed excess manganese (see the EDXS spectrum in Fig. 5.9). The microhardness of the material varied from 90 HRB to from 20 to 24 HRC at different locations.

Solution. The variation in microhardness showed that the material had hard and soft spots because of the black precipitates containing excess manganese. The precipitates were actually manganese sulfide added to the steel to increase machinability. The screw manufacturer was using the free-machining version of plain-carbon steel. The tensile strength of this steel is lower, thereby leading to premature failure.

Recommendation. By switching to standard grade 1020 steel, which has a higher tensile strength than the machinable version, premature failure was eliminated.

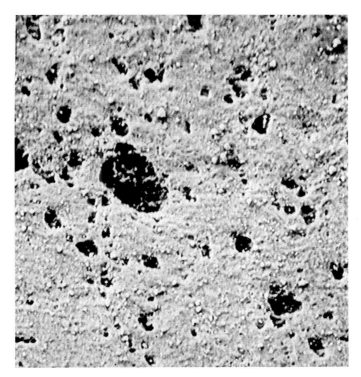

Fig. 5.8 SEM micrograph of a failed screw showing black spots and variable microhardness.

CASE STUDY 5.7: Aggressive Fluxing of Copper in Flex Circuits

Problem. One lot of flex circuits exhibited an abnormally high failure rate of the copper lead at the soldering point during manufacturing and assembly.

Analysis. The flex circuit with the broken copper strand was enclosed in a plastic mold and polished to obtain the cross section through the center of the strand at the fractured area (Fig. 5.10). Fracture was observed at the site of the copper strand exposed through a small hole in the polyimide outer cover of the flex circuits. These exposed copper areas were covered with a solder material that was applied by the vendor in a presoldering operation. The solder

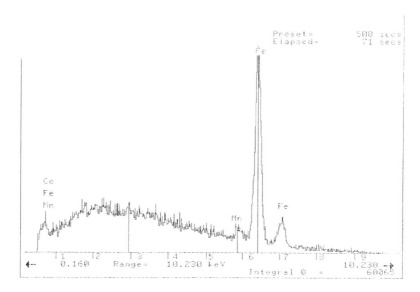

Fig. 5.9 EDXS spectrum of the black spot shown in Fig. 5.8, indicating the presence of excess manganese from manganese sulfide added for increased machinability.

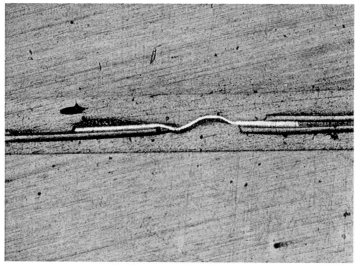

Fig. 5.10 Cross section of a fractured flex circuit showing the semicircular copper area exposed through a small hole in the polyimide cover. Solder material was chemically etched away. 7.8×.

Fig. 5.11 Exposed area of flex circuit thinned approximately 50% by over-etching. Note that the over-etched area is very rough and thus susceptible to unexpected fracture during assembly. 156x.

material was chemically etched away to reveal the copper substrate surface.

Figure 5.11 shows that the copper at the exposed area was thinned by about 50% (compared to the original thickness of the strand under the polyimide cover) during presoldering, and the copper surface is very rough.

Solution. During presoldering, the copper strand was etched to increase the adhesion with the solder material. Over-etching increased adhesion, but made the surface rough and the strand thin. As a result, the strand became fragile and cracks could initiate at the rough surface, leading to fracture during assembly.

Over-etching decreased the thickness of the copper strand by about 50% and made the surface very rough. When stress was applied on the flex circuit during assembly, cracks started at the stress points on the surface, propagated through the thickness, and led to fracture.

Recommendation. The vendor was advised to decrease the aggressiveness of etching (by decreasing the time of etching and the concentration of the etchant) prior to applying the soldering material.

CASE STUDY 5.8: Sensitization of Stainless Steel

Problem. A type 302 stainless steel piece experienced corrosion in a humid environment.

Analysis. Stainless steel should not exhibit any corrosion in humid environments. The composition of the material was verified with atomic emission spectroscopy and was found to be the same as the composition of type 300 series stainless steel. The piece was mounted in a plastic mold and polished to study its microstructure. Figure 5.12 is the optical micrograph of the sample after it was etched in an oxalic acid electrolytic etch. It showed the presence of dark precipitates at the grain boundaries. Figure 5.13 shows the microstructure of annealed type 300 series stainless steel.

Solution. The presence of dark precipitates at the grain boundaries indicates sensitization of stainless steel, which leads to the precipitation of chromium carbide at the grain boundaries and consequent depletion of chromium from the grains. Once the chromium level goes below the critical level of 18% in the bulk, stainless steel corrodes in humid environments just like steel. Careful tracking

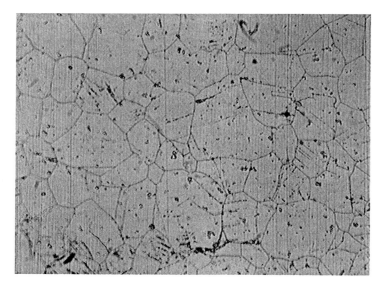

Fig. 5.12 Microstructure of corroded type 302 stainless steel showing evidence of sensitization and grain boundary precipitation of dark chromium carbide. 500×.

of the previous history of the stainless steel piece indicated that it was exposed to a temperature of about 650 °C (1200 °F), which is the sensitization temperature of stainless steel.

Recommendation. Stainless steel should not be exposed to temperatures in the range of 600 °C (1110 °F) for any length of time, because of the likelihood of sensitization occurring.

CASE STUDY 5.9: Failure by Fracture

Problem. An aluminum alloy sand-cast part experienced sudden failure by fracture.

Analysis. Figure 5.14 shows the fractured surface, indicating that the top portion failed in a ductile mode, but the bottom portion experienced brittle fracture due to the presence of blow holes and dirt near the edge.

Solution. The spot of dirt in the casting probably originated from molding sand particles or refractory slag entrapped by the molten metal. Blow holes or gas porosity may be caused by moisture or a volatile organic material. Both entrapped dirt and blow holes de-

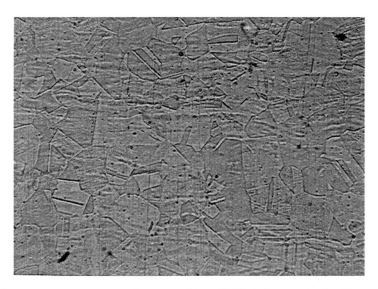

Fig. 5.13 Microstructure of annealed type 302 stainless steel without grain boundary precipitation. 500×.

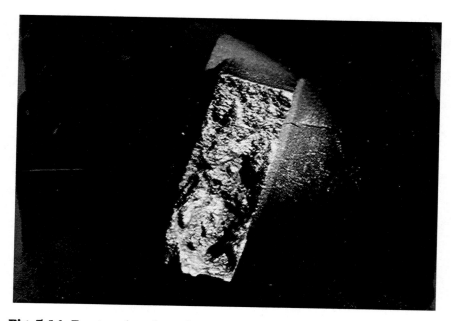

Fig. 5.14 Fractured surface of sand-cast aluminum part. Note dirt spot and blow holes in the bottom half of the casting, which led to fracture under load.

creased the strength of the casting, which fractured during load application.

Recommendation. Castings should be free of defects such as entrapped dirt, blow holes, or gas porosity.

CASE STUDY 5.10: Fatigue, Necking, and Arcing

Problem. A ground cable broke in service, and all of the copper strands fractured.

Analysis. The fractured surfaces of the copper strands were studied by SEM analysis and were categorized into three groups. Figure 5.15 represents the first group of the copper strands, in which a part of the fractured surface was smooth and the rest was rough. Figure 5.16 represents the second group, in which a part of the surface was smooth and the rest exhibited failure due to melting of the metal. Figure 5.17 represents the third group, in which all of the surface was rough with evidence of necking.

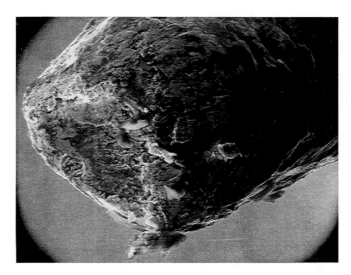

Fig. 5.15 Fractured surface revealing a smooth area, caused by brittle fatigue fracture, and a rough area, caused by ductile fracture. SEM, 390×.

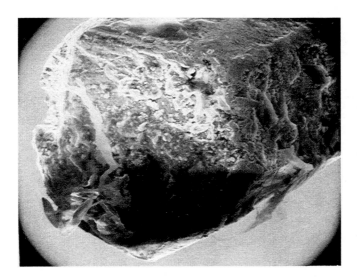

Fig. 5.16 Fractured surface revealing a smooth area caused by fatigue, and softening due to melting caused by arcing in neighboring strands. SEM, 390×.

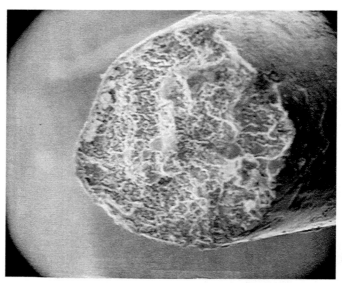

Fig. 5.17 Fracture surface exhibiting necking and complete rough surface caused by ductile fracture mode. SEM, 390×.

Solution. The first group of strands was fatigued due to repeated stress application. The metal became hard because of plastic deformation and broke in a brittle mode, creating a smooth fractured surface. Stress then became concentrated in the rest of the cross section of the strand, which broke in a ductile mode, creating a rough fractured surface.

The second group was fatigued to some extent and partly broke in a brittle mode. It appears that arcing occurred from the already broken strands, which heated the neighboring partially-broken strands to close to the melting point. When a metal becomes hot, it flows easily due to plastic deformation and fractures at a much lower stress.

The third group of strands was subjected to all of the load when the strands of the first and second groups broke. The former broke in a ductile mode after necking, thereby creating a rough fractured surface.

Recommendation. There was nothing wrong with the integrity of the strands and the strand material. The cable should be routed in such a manner so that the number of stress reversals on the strands is reduced.

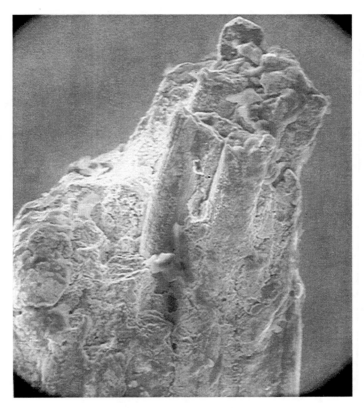

Fig. 5.18 Fractured cable showing that some strands broke at the solder joint and some a short distance from where the cable was bent over a prong. SEM, 250×.

CASE STUDY 5.11: Stress Concentration Leading to Fatigue

Problem. In a plug and cable assembly, one cable failed by fracture at the solder joint, and another failed a short distance from the soldered joint. Both failures were premature.

Analysis. Figure 5.18 shows that the fracture occurred at two different locations—one at the solder joint and the other a short distance away. Figure 5.19 is a micrograph of the fractured surface showing the smooth area (brittle fracture due to repeated stress reversals), the rough area (ductile fracture when it finally and com-

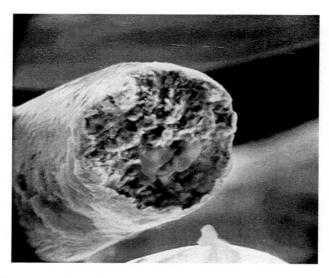

Fig. 5.19 Fractured surface showing a smooth area, caused by brittle fracture; a rough area, caused by ductile fracture; and a stressed area around the fracture. SEM, 780×.

pletely broke), and the stressed areas around the fracture zone. Figure 5.20 shows that the as-soldered joint was defect-free, because there were no cracks or other defects.

Solution. All fractured areas exhibited the same features. There were no defects in the cable material. Both smooth and rough areas were present on the fractured surface, indicating fatigue failure, and stressed areas were visible around the fractured area. The two failed cables were found to be shorter than other cables, and consequently they were stressed more. The solder joint was the stress concentrator for the cable that broke at the joint. In the case of the second cable, it was bent over a prong that acted as a stress concentrator. As a result, the cables were fatigued due to repeated stress reversals and fractured.

Recommendation. There was nothing wrong with the cable material and the soldering process. All cables should be of the same length. Any extra length of cable should be kept in the groove of the plug so as to absorb the shocks and stress reversals produced by the repeated stresses in the cable.

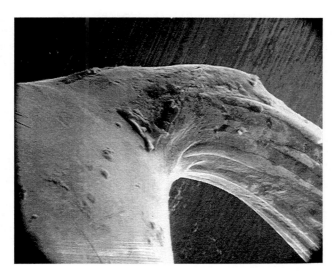

Fig. 5.20 As-soldered cable typical of a defect-free soldering process.

CASE STUDY 5.12: Bends as Stress Concentrators

Problem. A steel clutch spring exhibited frequent failure due to fracture in service.

Analysis. Figure 5.21 is a micrograph of the fractured surface, showing the smooth area (brittle fatigue fracture), the rough area (ductile fracture), and the stress concentration area.

Solution. There was nothing wrong with the spring material itself. It appears that the stress concentrating area led to hardening and crack initiation, and the spring failed due to normal fatigue failure.

Recommendation. The stress concentrating point was due to either a 30° bend at the end of the spring or a small bend radius. Removing the 30° bend and increasing the bend radius made the area smooth, removed the stress concentration, and eliminated the fracture problem.

CASE STUDY 5.13: Spring Failure Caused by a Sharp Bend

Problem. A spring with a 90° bend experienced premature failure. The spring was made of AISI 1095 high-carbon steel with high hardness.

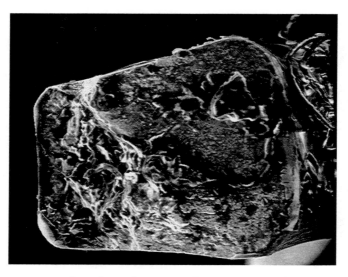

Fig. 5.21 Fractured surface of a spring showing a smooth area, caused by brittle fracture; a rough area, caused by ductile fracture; and the stress concentration area. SEM, 171.6×.

Analysis. Figure 5.22 shows the cross-sectional area of the spring near the 90° bend. In the horizontal section near the bend, the spring is bent slightly, thinned, and deformed. Die marks are also visible in the same area. Figure 5.23 shows that the fracture occurred in the deformed section of the spring. The spring was mounted in a plastic mold, polished, and etched. Figure 5.24 is a photomicrograph of the cross section, indicating that the crack started at the inside edge of the deformed section.

Solution. In the area next to the 90° bend, a part of the spring became thin due to insufficient metal flow. At the location where the die ended, a sharp indentation was observed in the spring, reducing its thickness and making it weak. During normal use of the spring, stress was concentrated in the deformed area, which was already weakened by the die marks and thickness reduction, leading to crack initiation. Because AISI 1095 steel is a very hard material, cracks propagated easily through the material, and the spring failed under cyclic load.

Recommendation. The 90° bend itself was defect-free, but the bending process created a defect in the spring. Because bend radius

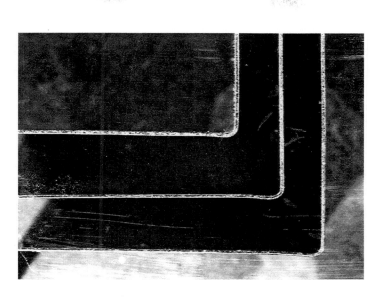

Fig. 5.22 Cross section of a spring with a 90° bend, showing the bent, thinned, and deformed section near the bend. 7.8×.

Fig. 5.23 Cross section of a spring showing that the fracture occurred in the deformed area near the bend. 15.6×.

Fig. 5.24 Cross section of a spring showing that fracture started at the inside edge in the deformed section. 156×.

was not critical in this application, it should be increased so that the stresses are distributed over a larger area, and stress concentration is reduced.

CASE STUDY 5.14: Mechanical Integrity of a Metallic Bond

Problem. A bronze leaf spring was soldered to a silver graphite stud in a static ground application. The part needed to be mechanically and metallurgically evaluated as part of the approval process.

Analysis. Microscopic study (at low magnification of 20× or under) revealed that the solder was even around the studs and there were no cracks in the silver graphite contact material. The contacts were tested in a tensile testing machine. The pull strength exceeded 4 kgf (8.8 lbf) when the contact material broke, indicating that the joint was stronger than the contact material itself. The soldered joint was encased in a plastic mold, polished, and etched. Figure 5.25 shows that the bond between the silver graphite and the bronze leaf was sound, with no voids or gaps at the joint. Hardness of the leaf spring at the center of the joint was found to be the same as that of the bulk leaf spring material, indicating that heating during soldering did not deteriorate the mechanical properties.

Fig. 5.25 Soldered joint of silver graphite and bronze leaf indicating a good bond quality without voids or gaps. 156×.

Solution. The solder bond quality was good, and the bond strength was higher than the tensile strength of the material itself. The soldering process did not affect the mechanical properties of either component.

Recommendation. The part should be approved for usage.

CASE STUDY 5.15: Surface Scratching on a Motor Shaft

Problem. A motor shaft failed prematurely due to fracture. The shaft was made of SAE 1040 or 1045 plain-carbon steel.

Analysis. Figure 5.26 shows the fractured surface, revealing the ductile fracture. However, around the outer edge of the circular cross section of the shaft, there was evidence that a sharp object dug into the shaft. Figure 5.27 shows the shaft surface close to the fractured surface, which was much rougher than the smooth machined surface of the shaft. This further confirms that a sharp object dug into the shaft. Figure 5.28 shows the microstructure of the shaft, indicating that it was made of annealed SAE 1040 or 1045 steel with ferrite and pearlite structures. Its hardness was found to vary between 24 and 27 HRC.

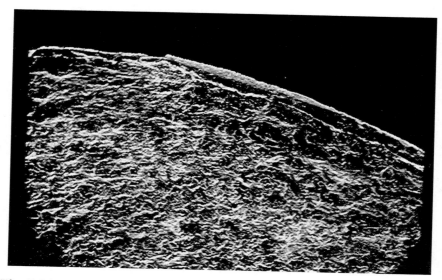

Fig. 5.26 Fractured surface of a shaft showing its cross section. SEM, 50×.

Fig. 5.27 Outer surface of the shaft showing roughening near the fracture site, caused by a set screw digging into its surface. SEM, 39×.

Fig. 5.28 Microstructure of the shaft showing white ferrite, dark pearlite, and the fully annealed SAE 1040 or 1045 structure. SEM, 395×.

Solution. The set screw that tightened the brake onto the shaft slipped and dug into the shaft, creating a notch around the periphery, which was a stress concentrator. Cracks formed at these stressed areas and propagated through the cross section of the shaft. As indicated by the low hardness, the shaft material was soft, thereby providing low resistance to fracture.

Recommendation. Clamping of the brake should be redesigned so as to avoid the potential problem of digging into the shaft. Also the shaft should be heat treated and tempered to 35 to 40 HRC to improve its hardness, ductility, and impact and fatigue properties.

CASE STUDY 5.16: Elastomer Hardening in Sulfur-Containing Atmospheres

Problem. A customer noticed that the elastomer drive belt was creating considerable powder and dust contamination. The belt was deemed a failure and returned. However, the same problem was not evident at other customer locations.

Analysis. Examination of the inner surface of the belt revealed that it was smooth and appeared to have frequently slipped off the pulley. Tensile strengths of the failed belt and an unused belt from the same lot were measured at 370 and 455 kgf (815 to 1000 lbf), respectively, whereas the specified tensile strength was 500 kgf (1100 lbf). Chemical analysis of the two belts indicated similar chemical compositions.

Solution. It was clear that the belt was becoming progressively harder with a correspondingly lower tensile strength during usage at that customer location. Belts obtained after a similar length of service at other locations did not exhibit a similar decrease in tensile strength. Neoprene-type elastomers are known to continue to cure in the presence of excess sulfur, which makes the surface hard, smooth, and polished, with very low friction. Consequently, the belt would slip from the pulley, generating particulate contamination due to abrasive wear. An investigation of the customer site confirmed the presence of excess sulfur in the atmosphere in which the drive was used.

Recommendation. The customer was informed as to the effect on mechanical properties of excess sulfur in the work atmosphere. If the drive cannot be shielded from the atmosphere, the belt should be made of a material that will not deteriorate in the presence of excess sulfur.

CASE STUDY 5.17: Effect of Heat Treatment on Fatigue Failure

Problem. A retaining rod made of type 410 stainless steel in a disk pack exhibited premature failure.

Analysis. Figure 5.29 shows the fractured surface, indicating a partly smooth surface (brittle fracture due to hardening), with the rest a rough surface (sudden ductile fracture). The fracture occurred where the rod was bent into the required shape.

Solution. The fracture surface study indicated that it was a typical fatigue fracture. The rod was highly stressed in the bend area and lost ductility due to cold working during bending. The bending process was not optimized because the operation left nicks and microcracks at the bend radius. These imperfections acted as stress concentrators, leading to fatigue failure.

Recommendation. The bending operation needed to be optimized so that it would not cause mechanical damage or nicks at the bend radius. Type 410 stainless steel loses all ductility by cold working 50%, and consequently, the rod should be stress relieved after bending by annealing at 320 °C (600 °F). If a material change is

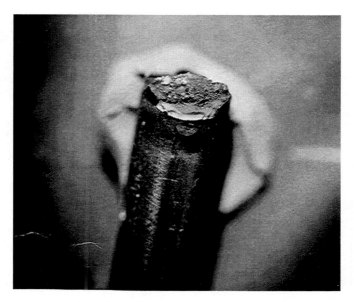

Fig. 5.29 Fractured surface of a retaining rod showing fatigue failure. Bending made the material hard and crack-prone. 10×.

possible, type 414 stainless steel should be used because of its higher impact strength.

CASE STUDY 5.18: Effect of Carbon Content on Corrosion Resistance of Stainless Steel

Problem. One lot of connector block pins exhibited a buildup of scale during normal usage. The humidity level was determined to be quite high in the work environment. The pins were made of type 416 stainless steel. The corroded and uncorroded pins were analyzed.

Analysis. The uncorroded and corroded pins were found to have hardness values of 22 and 43 HRC. Figures 5.30 and 5.31 show their respective microstructures.

Solution. The uncorroded pins contained less carbon than the others and fully annealed structure, and they did not exhibit corrosion buildup. The corroded pins had a higher carbon content and a fully hardened martensitic structure. Even though both lots were made of type 416 stainless steel, the corroded lot contained excess carbon. During the pin manufacturing process, the material was

Fig. 5.30 Microstructure of an uncorroded pin, showing annealed structure, lower carbon, lower hardness, and no corrosion. 156×.

Fig. 5.31 Microstructure of a corroded pin, showing hardened martensitic structure, higher carbon, higher hardness, more chromium carbide precipitation, and corrosion buildup. 390×.

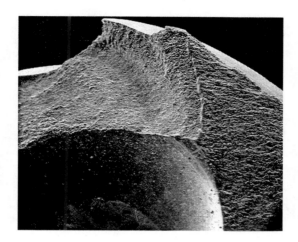

Fig. 5.32 Fractured surface of a lock screw, indicating that failure was caused by a fatigue fracture that started at the minor radius in the threaded area. SEM, 7.8×.

exposed to the sensitizing temperature of 550 to 600 °C (1020 to 1100 °F), which facilitated the formation of chromium carbides. As the chromium was tied up in these carbides, the rest of the material was depleted in chromium and thus corroded in a moist environment.

Recommendation. It is important to monitor the carbon content of stainless steel. The material should not be kept in the sensitizing temperature range during processing. Heating treating operations should be monitored so that the material does not pick up any carbon.

CASE STUDY 5.19: Improper Materials Selection for Lock Screws

Problem. One lot of lock screws made of type 440C stainless steel exhibited premature failure due to fracture.

Analysis. The failed screws and the working screws were analyzed. Figure 5.32 shows the fractured surface, which indicates that a fatigue fracture initiated at the minor radius in the threaded area.

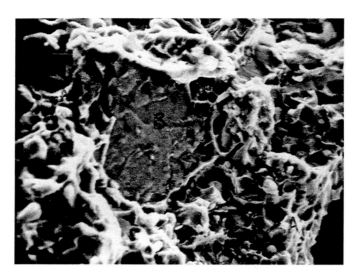

Fig. 5.33 Microvoids, microporosity, and dimples on the fractured surface near the area where crack propagation started. SEM, 1560×.

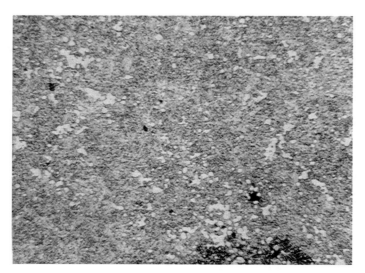

Fig. 5.34 Microstructure of screw material from the acceptable lot showing no porosity and an even distribution of small spheroids of carbide in a matrix of tempered martensite. 390×.

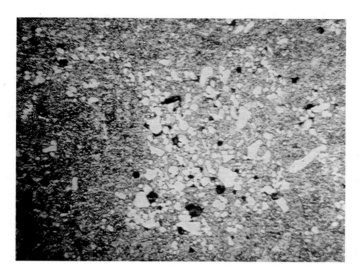

Fig. 5.35 Microstructure of the failed screw material showing microporosity and chromium carbide precipitation, indicating poor quality raw material and improper heat treatment, which caused sensitization. 390×.

Figure 5.33 shows dimple formation near the area where the crack started. The microstructures of the working and failed screws are shown in Fig. 5.34 and 5.35, respectively. In the working lot, the carbide particles were small spheroids evenly distributed in a matrix of tempered martensite, whereas in the failed lot, chromium carbide precipitation and microporosity were evident. The working lot had a hardness of 55 HRC, whereas the failed lot had hardness values ranging from 48 to 52 HRC at various locations.

Solution. A dimple in the microstructure is half of a microvoid through which fracture occurs. A microvoid is generally initiated at interfaces between the matrix and imperfections such as microporosity, microcracks, carbide precipitates, and inclusions. The microvoids grow under the triaxial stress conditions ahead of the crack tip and expand until they coalesce to leave behind hemispherical cavities known as dimples on the fracture surface. In this case, the fracture must have originated at the site of a materials defect, leading to the formation of a major crack from various microcrack nuclei. The crack then must have propagated by microvoid initiation, growth, and coalescence.

The variation in hardness values of the failed lot was due to materials imperfections such as micropores and carbide precipitates.

Recommendation. The failed lot contained materials defects such as micropores and voids. Additionally, the material must have been kept in the sensitizing temperature range, leading to chromium carbide precipitation. If stainless steel is sensitized, it should be solution annealed in the range of 1050 °C (1900 °F) where all carbides would be dissolved. Then the material should be tempered at a given time and temperature to obtain the required hardness and structure of the small tempered spheroid carbides in the martensite matrix. After heat treatment, the material should contain less than 5% retained austenite.

Appendix 5.1

Use of Failure Analysis Methods
Improves Flex Circuit Reliability*

Flexible circuitry is commonly used to connect moving parts to a stationary circuit board. As its name indicates, it has excellent flexibility and is used in fast-moving applications. Because the flex circuit has very low mass, it is used in load-sensitive applications. It can replace many wires and printed circuit boards and can stay out of the way of the moving parts.

The flex circuit generally consists of flat rolled copper conductors (about 150 μm wide and 25 μm thick) sandwiched between two thin polyimide sheets (e.g., Kapton from DuPont) with the help of an adhesive—typically an acrylic. The rolled copper conductors are generally acid or alkali etched to roughen the surface and remove mill scale and grease-type hydrocarbon contamination. Both surface roughening and surface cleaning improve the bonding of copper with the adhesive and the polyimide and ensure that the conductors and the polyimide sheets do not separate during the life of the flex circuit. Because the thin copper conductors have a very small cross section, they are easily damaged by even relatively small loads during handling.

During the assembly and operation of the device, frequent flexing of the circuit and consequent stress reversals subject the copper conductors to enormous stress. The latter can lead to breakage of the conductor, primarily due to fatigue and failure of the flex circuit. Often it is difficult to locate the break because the copper conductor

---

* P.B. Narayan and A.S. Brar, *Connection Technology*, Oct 1988, p 19-21. Reprinted with permission.

lies buried in the polyimide material. However, the fractured surfaces of the conductor must be studied in an undamaged condition to analyze the failure and eliminate the causes. The following is a case study of a flex circuit failure using a technique to remove the polyimide and the adhesive, exposing the copper for examination. Included in the case study is information on choosing the right material for the copper conductor with an optimum surface finish, because the fatigue resistance of the conductor depends on surface finish, stress reversal rate, maximum stress, and other factors.

A flex circuit with a breakage in one of the conductors was examined with an optical microscope, and a dark region was discovered where the conductor could possibly have a fracture. Scanning electron microscopy (SEM) could not provide information, because it is a surface technique and the secondary electrons (which form the

Fig. 1 SEM micrograph of flex circuit after plasma etching. The conductor on the left exhibits fracture. Inset depicts fracture magnified.

SEM image) cannot penetrate the polyimide layer. Cross-sectioning by mounting the specimen in a resin mold and polishing was attempted, but is very difficult to look at a fractured surface undamaged by this technique. Thus, a technique was necessary to remove the polyimide from the surface and expose the fractured surface of copper with all its microstructural features intact. Low-temperature plasma, which consists of neutral, energized, and ionized atoms of a gas produced by an electric field, can be effectively used for removing polyimide and the adhesive from the surface. The plasma is generally very reactive with organic materials and almost passive with inorganics, metals, and ceramics. It is mainly a surface technique because the plasma affects only the top few hundred angstroms. It does not affect the bulk properties. It is called a low-temperature plasma because the temperature increase of the specimen due to exposure is about 30 or 40 °C (54 to 72 °F) above ambient. Many gases, including oxygen, argon, nitrogen, hydrogen, and carbon tetrafluoride and their mixtures in varying proportions, can be used for generating plasma. In addition to bias voltage, etching time, and other variables, the most important parameter of plasma etching is the optimum gas mixture.

Thus, using low-temperature plasma, the copper conductors and the fractured surfaces were exposed in the undamaged condition. The SEM was used for surface topographical studies of the fractured surfaces to determine the failure mechanism.

Failure Analysis

The location, where the failed flex circuit was used, was closely studied. During use, the flex circuit was firmly attached to a block made of sand-cast aluminum with a metal clip. The cast block showed sharp protrusions on the surface at a point corresponding to the location of the riser in the cast mold.

Figure 1 shows a SEM micrograph of the flex circuit soon after plasma etching. The left conductor exhibits fracture. The inset in Fig. 1 shows the same fracture isolated and at a higher magnification. Figure 2(a) shows the fractured surfaces, and Fig. 2(b) and (c) show that the cracks existed at the surface along the breadth of the conductor. These illustrations indicate that the cracks propagated along the thickness during the flexing caused by assembly and operation of the device, and the circuit failed long before the expected fatigue life of the conductor.

(a)

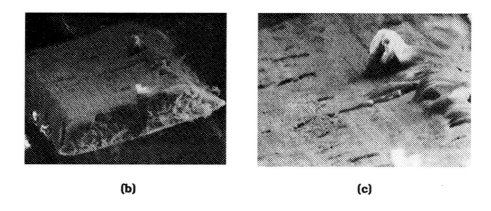

(b)   **(c)**

Fig. 2   (a) Fractured surface of the conductor. (b) and (c) Cracks on the surface of the conductor.

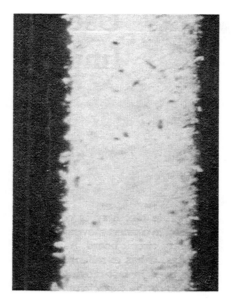

Fig. 3 Cross section of a copper conductor showing roughness of the surface.

From this analysis, it can be surmised that the flexing surface is unduly rough due to excessive chemical etching and cleaning. Because flexing involves many stress reversals, it could permit the forming of a microcrack in the weakened area and propagate that crack through the conductors, leading to premature failure. The sharp protrusions on the casting and the holding clip act as stress concentrators, thus contributing to the crack initiation.

Materials Selection

The copper conductors display a very rough surface on both sides, as may be seen from the cross section of a conductor in Fig. 3. The chemical etching of copper appears to be excessively harsh and severe, making the surface unduly rough, which generates weak areas. The harshly etched surface, surface roughness, and weak areas can be seen in Fig. 4.

Figure 6 shows a cross section of the conductor with one side mildly etched (Fig. 6a) and the other side as rolled (Fig. 6b). Because rolling introduces beneficial compressive stresses, the conductor will

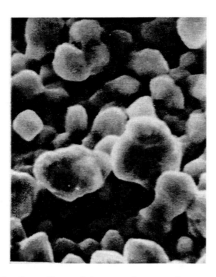

Fig. 4 Harshly etched surface of the conductor showing surface roughness and weak areas.

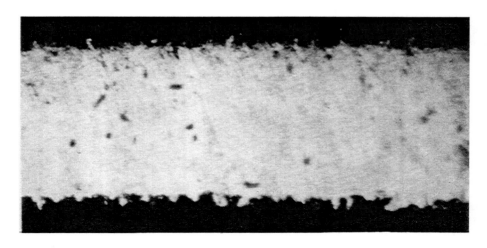

Fig. 5 Cross section of the conductor with one side mildly etched and the
other side as rolled.

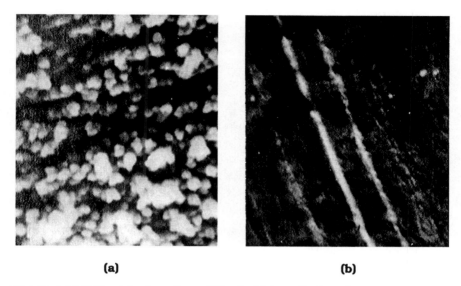

(a) **(b)**

Fig. 6 (a) Mildly etched surface of Fig. 5. (b) As-rolled surface.

be less prone to cracking and can be expected to have a significantly longer fatigue life than the conductor that was harshly etched on both sides. The surface roughness profiles for the severely etched, mildly etched, and as-rolled conductors are shown in Fig. 7. The mildly etched conductor may exhibit less adhesion, but the most important advantage is that it is not prone to cracking and premature failure during flexing. Very strong adhesion is not required in this application, and the mildly etched conductor has enough adhesive strength.

Conclusion

In a flex circuit, copper conductors generally fail due to fatigue. Severe etching during cleaning—prior to bonding—reduces fatigue life by introducing weak areas and stress concentrators on the surface, which should lead to premature failure during stress reversals involved in flexing. Consequently, the conductors should be mildly etched.

It is sufficient to etch copper only on one side to reduce the chance of generating weak areas and stress concentrators. A portion of the

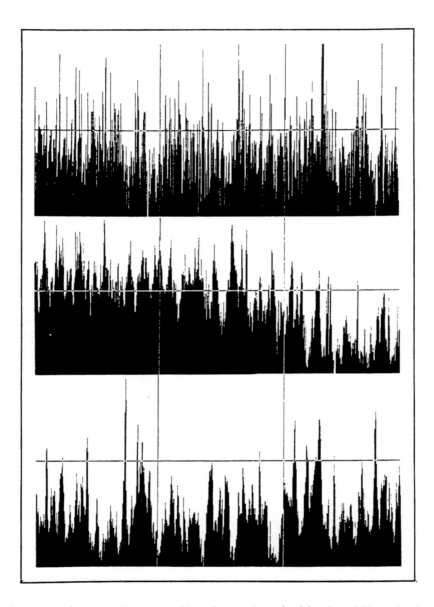

Fig. 7 Surface roughness profiles of severely etched (top), mildly etched (center), and as-rolled (bottom) surfaces.

adhesion strength between copper and polyimide may have to be sacrificed in this regard. Better cleaning and roughening techniques should be studied for preparing copper conductors for flex circuits. A promising technique is low-temperature gaseous plasma cleaning using a suitable gas mixture.

A convenient method of analyzing flex circuit failure is to remove polyimide with plasma ashing in a suitable gas mixture. Low-temperature plasma will not disturb the fractured copper surface. Optimum design of the parts used in conjunction with flex circuits should avoid sharp protrusions and edges, which act as stress concentrators.

The authors acknowledge with pleasure the help and advice of Mr. Bernie Rapp of Branson/IPC during the plasma treatment.

Chapter 6

Microcontamination

Three types of inductive read/write heads can be used in hard disk drives. In low-cost drives, monolithic heads are used, in which a nickel-zinc ferrite closure is glued onto a nickel-zinc ferrite slider. For improved performance, composite heads are used, in which a manganese-zinc ferrite core is encased in a ceramic head pad material such as calcium titanate. In high-performance drives, thin-film heads are used, in which magnetic poles are deposited at the end of a composite ceramic (e.g., alumina-titanium carbide). Future drives will contain dual heads, manufactured using thin-film head technology, with inductive write and magnetoresistive read elements. See Appendix 6.1 for more information on disk drive manufacturing and technology.

Information is stored on a disk, in which a thin layer of magnetic material is deposited on a polished aluminum substrate. The magnetic material consists of a 400 nm (15.8 μin.) thick epoxy layer containing particulate gamma iron oxide or 30 nm (1.18 μin.) thick cobalt alloy, plated or sputtered onto the polished aluminum substrate. Consequently, the disk is softer than the head.

In a Winchester-type disk drive, the head rests on the disk when the drive is turned off. When the drive is turned on, the disk begins to spin, assuming that the breakaway force is greater than the static friction (stiction) between the disk and the head. Because dynamic friction is lower than stiction, the frictional force decreases once the disk starts spinning. When the disk reaches a certain velocity, the head becomes airborne, supported by the air bearing generated by the spinning disk. The disk is designed to reach a specified rotational speed (3600, 5400, or 7200 rpm), at which point the head is "flying" at a specified height, e.g., 100 to 200 nm (3.9 to 7.9 μin.) from the disk. When the drive is turned off, the head lands on the disk.

6.1 Importance of Stiction

Stiction is one of the most important factors that affect drive reliability (Ref 1). If the stiction is too high, the head breakaway process will be quite violent and will permanently damage the fragile head suspension mechanism. Sometimes stiction is so high that the head peels off some of the magnetic medium from the disk during the breakaway process. Any such damage could lead to loss of information and a catastrophic "head crash," in which the flying head accidentally touches the disk and scrapes off the magnetic medium.

Stiction can be reduced by (1) changing the head design, e.g., using a smaller slider, or positive crown, etc., (2) reducing the area of contact at the head-disk interface by increasing the disk surface roughness, and (3) reducing moisture and eliminating hydrocarbon contamination at the interface.

Because the disks and heads are within an enclosed environment, contamination generated by any drive component accumulates at the head-disk interface. It is thus imperative that none of the drive components generate any type of contamination.

6.2 Sources and Types of Microcontamination

The primary sources of microcontamination are the atmosphere, human handling, and the drive components. Contamination can be classified as particulate, magnetic, and film-like.

6.2.1 *Particulate Contamination*

When particles are large, they can impinge on the disk and scratch it. If particles are smaller than the head flying height, they can accumulate at the head-disk interface, thus causing signal degradation and possibly a head crash. These small particles can cause the most significant damage (Ref 2).

Atmospheric silicate-type dust particles, wall and ceiling construction materials, and carpet fibers are common sources of particulate contamination. Human handling produces cloth fibers, flakes of skin, traces of cosmetics, pieces of hair, glove prints, fingerprints, and spittle marks. Appendix 6.2 describes spittle contamination in detail.

Drive components can generate particles by (1) adsorbing particles from previous processing steps (e.g., machining), (2) flaking of the protective coatings, (3) oxidation leading to the generation of loose oxide particles, (4) corrosion of metallic components, (5) accumulation of wear debris at mating surfaces, and (6) condensation of airborne hydrocarbons.

6.2.2 *Magnetic Contamination*

Magnetic contamination constitutes a particulate-type phenomenon. However, it is extremely detrimental to drive performance and deserves special emphasis. If the magnetic particles accumulate on the head, they act as magnets that erase the data written on the disk. Under optimum conditions, the disk drive manufacturing area should contain only two magnets—a hard magnet, or the disk, and a soft magnet, or the head.

Some of the common sources of magnetic particles are memo and desk magnets, telephones, radios, and cassette players. Their use in and around clean rooms should be strictly controlled, if not prohibited. Various testing devices contain magnets. Table 6.1 lists some typical magnetic materials in disk drives, per Ref 3.

6.2.3 *Film-Like Contamination*

Film-like contamination is the most insidious type of contamination in that it is difficult to detect. Hydrocarbons become airborne, are transported in vapor form, and are deposited in liquid and solid form at the head-disk interface, causing a significant increase in stiction. When film-like contamination is present on components,

Table 6.1 Typical magnetic materials used in disk drives. Source: Ref 3

Material	Coercivity A/m (Oe)	Application
Samarium cobalt	8250 to 10,000 (6600 to 8000)	Voice coils
Strontium ferrite	4750 (3800)	Voice coils
Barium ferrite	1250 to 3500 (1800 to 2800)	Spindle motors
Neodymium-boron-iron	11,250 (9000)	Spindle motors
Nickel-zinc ferrite	0.19 (0.15)	Read/Write heads
Manganese-zinc ferrite	0.13 (0.10)	Read/Write heads

loose particles from previous processes easily adhere to their surfaces.

The machining environment contains oils, and human handling contributes body oils. Porous surfaces can entrap cleaning solutions and other fluids and introduce those hydrocarbons into the drive environment. Cast drive components can give off mold release agents. Partially cured adhesives and resins have the potential to outgas volatile components. All gaskets and rubber and vinyl parts may outgas, thus giving off volatile constituents.

Desirable features of plastics and adhesives to be used in drives include (1) low-temperature curing and quick setting, (2) no outgassing, (3) no particle generation, (4) no extractables, (5) low or no nonvolatile residues, (6) high bond strength, and (7) resistance to solvents, humidity, and temperature.

6.3 Evaluation of Cleanliness

Every drive component should be tested as a potential source of contamination. Some of the typical testing techniques used to evaluate drive components are discussed below.

6.3.1 *Turbidity*

Turbidity testing is one method by which the level of particulate contamination of a part surface can be quantified. This test procedure is based on ASTM specification D-1889-81, which describes a method for determining the particulate contamination level in a water sample through the use of the light-scattering characteristics of the particulate matter.

The part is washed in deionized water to which a small amount of surfactant (typically two drops to a liter of water) is added. The solution is agitated ultrasonically for about 2 min. The surfactant facilitates detachment of particulates from the surface of the part into the solution. About 30 mL (1.8 in.^{3}) of solution is used for every 11.6 mm^2 (square inch) of part surface area.

A beam of light is passed through the solution after the washing procedure is complete. Any suspended particles in the solution scatter the light. The intensity of the scattered light is compared to a standard formazin suspension and is given an index number. This empirical number becomes a repeatable standard for comparing particulate suspensions between solutions.

Depending on the location and criticality of the part, the design engineers must determine the level of cleanliness required and must specify the acceptable index number. To ensure accurate results, the surface area must be measured to within ±5%.

This method cannot indicate particulate size distribution or chemical composition. To obtain this information, it is necessary to filter out the particulates and perform image analysis and energy-dispersive X-ray spectroscopy (EDXS).

6.3.2 *Densitometer Testing*

This technique is useful as a semiquantitative method for specifying cleanliness, especially from the standpoint of particulate contamination. In this test, a transparent tape (or a permanent mending tape) approximately 5 cm (2 in.) long is pressed onto the surface of the part with adequate pressure to ensure complete and uniform adhesion. The tape is removed in one smooth and uniform motion and placed on a clean glass slide in a densitometer.

In the densitometer, a visible light beam is passed through the tape. By measuring the intensity of the transmitted light, the optical density of the tape can be calculated. If the tape contains particulates from the surface of the part, more light is absorbed and scattered by the particulates, thereby registering a higher optical density value. In other words, the optical density is an indication of the amount of particulate contamination emanating from the part. This technique also does not provide particulate size distribution or chemical composition.

6.3.3 *Surface Compositional Analysis Techniques*

The surface-sensitive techniques (such as ESCA, AES, and FTIR) provide the chemical composition of very thin layers (a few nanometers thick) of hydrocarbons on the surface. Equipment cost is high, and the area of the part that can be analyzed is small. For a part with dimensions larger than 20 cm (8 in.), sample preparation could be destructive. These techniques are not well-suited for the production environment. They are discussed in detail in Chapter 2.

6.3.4 *Optically Stimulated Electron Emission*

With this technique, the surface of the part is exposed to high-energy ultraviolet (UV) radiation (Ref 4). The UV energy recombines with photons in the surface monolayer. The photons collide with the electrons, transfer their energy, and cause electron emission, which gives rise to an electronic current measured in picoamperes. The emitted electrons are collected by an electrode that is spatially separated from the surface. Any monolayer contaminant film attenuates the current measured and thus can be quantified.

Electron emission testing is a non-contact, non-destructive technique that is capable of detecting oxides and contamination layers a fraction of a nanometer thick. However, its spatial resolution is not very good because the scanned area is approximately 5 by 5 cm (2 by 2 in.).

6.3.5 *Thermal Analysis*

The outgassing characteristics of resins, adhesives, gaskets, and other polymeric materials are best evaluated by thermogravimetric analysis. The various types of thermal analysis techniques are described in Appendix 1.1.

6.3.6 *Contact Angle Measurement*

This method can be used to detect the presence of film-like hydrocarbons on the surface of parts. It also provides an indication of wettability. A drop of water is placed on the surface of the part, and the contact angle made by the drop with the surface is measured by a contact angle goniometer. The cleaner the surface is, the lower the contact angle. For example, a cast part containing mold release contamination exhibited a contact angle of 80°, indicating poor wettability. When the part was cleaned in low-temperature oxygen plasma, the contact angle was almost zero, indicating removal of the hydrocarbon contamination.

6.4 Surface Cleaning Techniques

6.4.1 *Solvent Cleaning*

Organic solvents can be used to dissolve hydrocarbon contamination and reduce adhesive forces that facilitate particulate attachment to the surface of a part. Solvent cleaning processes can be divided into two categories: cold cleaning and vapor degreasing.

6.4.1.1 Cold Cleaning

Cold cleaning methodologies use liquid aliphatic petroleum, chlorinated or fluorinated hydrocarbons, or a suitable mixture of solvents at room temperature or slightly elevated temperatures. Parts can be immersed and soaked in the solvent, or it may be sprayed or wiped on their surfaces. To effectively remove particulates and clean deeply recessed and inaccessible areas, a convenient form of agitation (e.g., ultrasonics or power jet spray) is added during immersion. This type of cleaning process is relatively inexpensive and can be easily automated using conveyor belts. The choice of solvent depends on the predominant form of contamination to be removed.

6.4.1.2 Vapor Degreasing

The vapor degreasing method uses the hot vapor of a chlorinated or fluorinated hydrocarbon solvent to clean the part. The hot vapor is generated by heating the solvent, which has higher mobility and thus better cleaning capability as its temperature increases. The vapor is often captured, condensed, and recycled.

Chlorofluorocarbons have enjoyed wide use in the various solvent cleaning processes because they are stable, nontoxic, unreactive with metals, and excellent solvents for the common organics. However, when the chlorofluorocarbons are released into the atmosphere, they migrate into its upper layers because of their stability, where they react with ozone, thereby depleting the ozone layer. Ozone absorbs the carcinogenic ultraviolet radiation from the sun, and protects life on the Earth from them. Most countries have signed the Montreal Protocol, which drastically decreases and ultimately will eliminate the production and use of chlorofluorocarbons. Hence, there is an increasing urgency to develop alternate cleaning processes, the most convenient being aqueous cleaning.

6.4.2 *Aqueous Cleaning*

Aqueous cleaning is the most widely used of all the cleaning techniques. Some of its features are discussed below.

6.4.2.1 Suitability to All Materials

Aqueous cleaning can be used for almost all materials, including plastics. However, plastics may become deformed at higher temperatures during oven drying. Drying at lower temperatures takes longer. Aqueous cleaning may not be suitable for special components such as finished bearings, in which the bearing grease can absorb water and modify its lubrication characteristics.

6.4.2.2 Dissolution of Coatings in Water

Chromate conversion coatings have been used as corrosion-resistant overcoatings for aluminum, magnesium, and steel components. These coatings dissolve in water and may not be suitable for aqueous cleaning unless they have an organic overcoating. Water reacts with sputtered alumina layers, forming soft and porous pseudo-boehmite (hydrated alumina), with inferior tribological properties.

6.4.2.3 Ultrasonic and Aqueous Jet Cleaning

Turbidity tests, along with particle counter tests, indicate that both ultrasonic agitation and water jet cleaning are quite effective in removing particles from surfaces. Ultrasonic agitation with deionized water also efficiently removes film-like hydrocarbon contamination. However, ultrasonic cleaning cannot be used on components such as printed circuit boards where it might damage IC chips and open soldered joints. It could also cause delamination in multi-layered structures. For aqueous jet cleaning, the water pressure is kept below 2.4 MPa (350 psi). Above that pressure, water erosion damage has been observed on aluminum and magnesium parts, and dimensional distortion of a delicate assembly (e.g., head-suspension) was reported.

6.4.2.4 Surfactants

Although surfactants aid in decreasing adhesion of particles to a surface, their presence is not needed for effective cleaning. Moreover, they tend to leave a monolayer film on the cleaned surfaces.

6.4.2.5 Cleaning Water Temperature

Cleaning efficiency does not depend on water temperature. However, the water temperature is kept at approximately 70 °C (160 °F) because it is much easier to heat the part from that temperature to the oven drying temperature than to heat it from room temperature.

6.4.2.6 Drying

The presence of retained water could cause corrosion of metallic parts. Consequently, it is important to schedule drying cycles following aqueous cleaning. First, the part is dried in an oven with an inert environment (e.g., nitrogen or argon). This gas should be devoid of any particulate contamination. Second, any water trapped in a blind hole or inside wire insulation can be removed by sending the parts through a vacuum dryer. To prevent localized freezing of water in the vacuum dryer, the parts are warmed by radiation heating.

6.4.2.7 Cost

Aqueous cleaning is relatively inexpensive. Water is readily available and plentiful. Because it is not a hazardous procedure, no special precautions need to be taken. However, because deionized water can dissolve metal ions, the waste water from this process may require special treatment before disposal. For large systems, recycling is more economical.

6.4.2.8 Special Problems with Thin-Film Read/Write Heads

Read/write heads go through extensive processing and must have a clean bearing surface at the head-disk interface. The thin-film heads have multi-layered thin films of sputtered alumina, sputtered NiFe, and plated NiFe at the surface of the air bearing.

Film Dissolution in Deionized Water. Alumina dissolves slowly in deionized water.

Boehmite Formation. Alumina reacts with water, forming a soft and porous hydrated alumina, called pseudo-boehmite, which causes reduced adhesion with the next layer and changes tribological characteristics at the head-disk interface.

6.4.3 *Electrocleaning*

In this process, the part is suspended in an electrolyte and is made either the cathode or the anode of an electrochemical cell, although cathodic cleaning, in which it is made the cathode, is more common than anodic cleaning. A low-voltage current, usually 6 to 12 V, is applied. Typically, the electrolyte is an alkaline solution. Because the part must act as a conductor, this process works primarily for metals. It is common practice to reverse the current during the last 5 to 10 s of the cathodic cleaning operation. After the cleaning operation, the part is rinsed in hot or warm water, dipped in an acidic solution to neutralize any residual alkali, and finally rinsed in cold water.

6.4.4 *Electropolishing*

Electropolishing also works for conductors, such as metals, only. The part is made the anode, with a copper cathode in an acidic or alkaline electrolyte, using a direct current at a relatively low voltage of 9 to 24 V, with an amperage of 540 to 5,400 A/m^2 (50 to 500 A/ft^2). This process dissolves the surface layers of the part, thereby removing all surface contamination. At the same time, any surface microroughness is removed, producing a smooth surface.

Electropolishing is a very effective deburring technique. At the location of a burr, current density increases, facilitating metal dissolution. To obtain uniform metal dissolution rates at various points on a part, current distribution should be maintained uniformly over the surface of the part. The dissolved metal reacts with the electrolyte, forming salts and sometimes precipitating out insoluble salts, which alters the bath composition. Good chemical control of the bath is required to achieve uniform electropolishing.

6.4.5 *Low-Temperature Plasma Cleaning*

Plasma is generated by ionizing a gas with an electric field. The energized ions interact with hydrocarbon contamination and either physically remove it or chemically convert it into gaseous products that escape into the atmosphere. Plasma can be made reactive with hydrocarbon contamination by adding oxygen, fluorocarbon (carbon tetrafluoride), etc., to the gas mixture. Metal oxides can be removed by adding a reducing species such as hydrogen to the gas mixture.

Plasma affects only the surface of a part, to a depth of only a few micrometers. Consequently, it will not affect its bulk properties. Because the part is rarely heated above 50 °C (120 °F), this type of procedure is called low-temperature plasma cleaning. Plasma can change the molecular weight of the surface layer, primarily organic materials, by scissioning (reducing the length of its molecules), branching, and cross-linking.

Scissioning. The energized ions in the plasma sometimes react with the long molecular chains on the substrate, breaking them into smaller molecular chains that are more volatile and come out of the surface. The presence of oxidizing species such as oxygen in the plasma facilitates this break-up process, which is the primary cleaning mechanism.

Branching. The energy supplied by the plasma constituents could knock out an atom in the linear chain of the organic substance and cause the chain to branch out by reacting with another organic molecule. The branched-out polymer may have quite different properties compared with the original chain.

Cross-Linking. The energized ions of the plasma sometimes encourage cross-linking among the long molecular chains forming a network structure. Cross-linking makes the polymer harder, tougher, and more adherent to the substrate, and this may not be desirable when the primary motive is surface cleaning. It is important to ensure that the composition of the plasma-generating gas is compatible with the organics of the substrate and the reaction desired.

Plasma cleaning can be used for all materials. Important process variables include the vacuum level in the chamber, the flow rate of the gas into the chamber, the energy applied to excite the gas, the processing time for the part in the plasma, and the type of gas or gas mixture used. Because the exhaust is primarily CO_2 and organics, this technique has no atmospheric disposal problems, which makes it very attractive. The atomically clean surfaces it produces make it an excellent surface preparation technique for bonding, medical applications, etc. Because plasma cleaning is a vacuum technique, its initial capital costs are high, although maintenance costs are fairly low.

6.4.6 *Solid CO_2 Cold Jet Cleaning*

This cleaning process involves blasting the part with dry ice pellets, particles of solidified carbon dioxide gas, with air or an inert gas as a propellent. The pellets disintegrate on impact, minimizing abrasion and tensile length deterioration of the surface of the part. The transformation of solid CO_2 to gas causes a thermal shock that severs the bond between the contamination and the substrate, allowing easy removal by conventional cleaning methods such as aqueous cleaning. Because the gas is nontoxic, disposal of the spent gas is not a problem. Because dry ice pellets are nonconducting, electrical parts can be cleaned by this method.

The cold jet process is controlled by three variables: velocity of the blasting media, mass density flow, and thermal shock. By regulating pressure, subsonic, sonic, and supersonic velocities can be obtained. The mass of the pellets can be adjusted by regulating the feed rate. Volumetric flow is controlled by using variable compression settings.

6.5 Case Studies

CASE STUDY 6.1: Bearing Grease

Problem. Some drives exhibited positioning errors. The prime suspect was the linear actuator.

Analysis. Close observation of the actuator revealed organic buildup on the rails on which the actuator moved. The drive was dismantled, and the organic contamination was collected by careful scraping. An FTIR spectrum of the material matched the spectrum of the grease used in the actuator bearings. Examination of the bearings revealed the presence of an excessive amount of grease.

Solution. The grease in the bearings was so excessive it leaked into the drive environment and was deposited on the flat actuator rails, interfering with smooth movement of the actuator, which led to positioning errors.

Recommendation. A new specification was formulated for minimum and maximum amounts of grease to be incorporated in the bearings.

CASE STUDY 6.2: Moisture Entrapment by Intermetallics

Problem. A critical drive component made of an aluminum alloy developed powdery spots on its surface after several weeks of storage.

Analysis. Figure 6.1 is an SEM micrograph of the corroded area. Note the porosity around the central precipitate. Numerous precipitates were found on the surface. Figure 6.2 is the EDXS spectrum of the precipitate, which shows aluminum, iron, and silicon.

Discussion. The EDXS data indicated that the precipitate was an AlFeSi intermetallic compound that was very hard, brittle, and chemically inert. Because the matrix alloy was much softer than the intermetallic, porosity developed around the intermetallic during surface finishing. The pores retained cleaning solutions and moisture, leading to corrosion during storage.

Solution. Intermetallic compounds can be etched by strong acidic dips, which can cause surface porosity that traps moisture and cleaning solutions. Elaborate and costly cleaning and drying techniques may then be needed to inhibit corrosion. A better solution is to use a purer aluminum alloy that contains very fine or no intermetallic compounds. Because the surface is more homogeneous in a purer alloy, it can be passivated relatively easily.

CASE STUDY 6.3: Silicon on Permalloy

Problem. Thin-film heads have permalloy 82% Ni and 18% Fe poles enclosed in sputtered alumina on a composite ceramic. These heads are lapped and polished to provide a flat, smooth surface. The permalloy pole faces were found to be damaged after surface finishing.

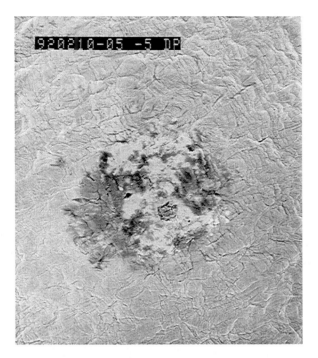

Fig. 6.1 SEM micrograph of corrosion on an impure aluminum alloy with a central precipitate.

Analysis. Auger electron spectroscopy analysis of the permalloy surface indicated the presence of silicon. Examination of the lapping procedure revealed that final polishing was done on a glass plate.

Solution. Permalloy was the softest of the head constituent materials exposed to the lapping and polishing procedure. Polishing debris from the glass (silicon dioxide) plate became embedded in the permalloy.

Recommendation. The final glass-lapping plate was replaced with a more wear-resistant lapping plate, which provided a defect-free permalloy surface.

CASE STUDY 6.4: Hydrocarbon Condensation

Problem. The copper shunt used to focus the magnetic flux in an actuator of a disk drive was found to be covered with yellow flakes.

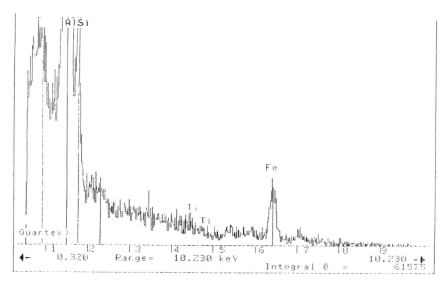

Fig. 6.2 EDXS spectrum of the central precipitate, which is an AlFeSi-type intermetallic, showing aluminum, silicon, and iron.

(a) (b)

Fig. 6.3 Yellow powder collected on copper plates. The gasket material outgassed and the airborne contaminant condensed on copper. (a) Low-magnification SEM. (b) High-magnification SEM.

Analysis. An SEM examination (Fig. 6.3) of the copper shunt revealed the hydrocarbon contamination on its surface, and the presence of hydrocarbons was confirmed by EDXS analysis. All plastic materials used in the drive were tested for outgassing. Thermal analysis revealed that the gasket material gave off hydrocarbons.

Solution. Copper acts as a catalyst for the condensation of hydrocarbons. Although the copper shunt was located some distance from the gasket, the gasket outgassed hydrocarbons that entered the drive environment and were deposited on the shunt because of the catalytic nature of copper. The copper shunt itself was not defective. In fact, it indicated the presence of film-like contamination in the drive environment.

Recommendation. The gasket material was replaced with a material that outgasses minimally under the given pressure and temperature conditions of operation.

CASE STUDY 6.5: Removal of Handling Contamination

Problem. Loose contaminants were found on a copper plate used in a disk drive in the as-received condition, which several cleaning

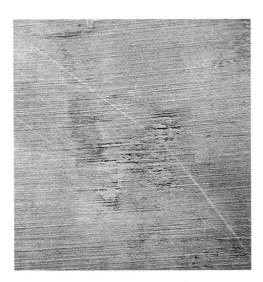

Fig. 6.4 Handling contamination (perspiration salts) pushed into the grooves of a machined copper plate. Contamination is difficult to remove once it occurs on plate.

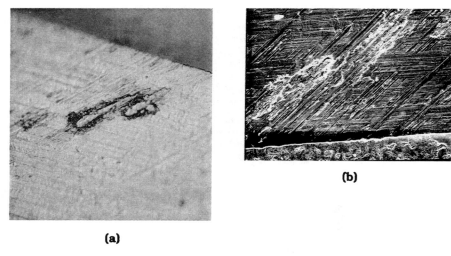

(b)

(a)

Fig. 6.5 Brown contamination on type 302 stainless steel caused by the reaction of smeared steel with retained water. (a) Optical, 96×. (b) SEM, 216×.

techniques, including mechanical removal and solvent cleaning, were unable to remove.

Analysis. Figure 6.4 shows SEM micrographs of the contamination. The presence of chlorine and sodium was indicated by EDXS analysis.

Solution. The presence of chlorine and sodium is indicative of human handling contamination (e.g., fingerprints). The salts are pushed into the fine grooves of the plate and are very difficult to remove.

Recommendation. The handling procedure was changed to require use of gloves. The easiest way to eliminate contamination is to prevent its occurrence. Cleaning the part later in the process is costly and time-consuming.

CASE STUDY 6.6: Contamination of Stainless Steel

Problem. A type 302 stainless steel part underwent a precision bonding operation and was attached to an aluminum part with a type 302 stainless steel screw. The screw was covered with a clear resin.

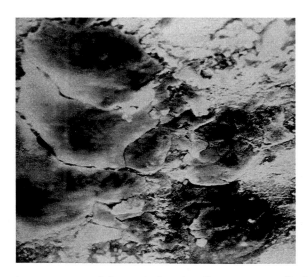

Fig. 6.6 Hard water mineral deposit showing the presence of calcium and small amounts of aluminum and silicon. SEM, 156×.

After several days, a brown stain appeared under the resin, even though the part was kept in a clean and dry environment.

Analysis. The brown stain was clearly visible when the resin was removed. Figure 6.5(a) shows optical micrographs of the area, and Fig. 6.5(b) shows SEM micrographs of the contamination. EDXS analysis of the area revealed the presence of iron, chromium, and nickel. However, the iron-chromium ratio was much higher than in the unstained area.

Solution. When the brown stain appeared, the presence of iron oxide was immediately suspected. The EDXS studies confirmed this, because the chromium and nickel peaks were much shorter than those emanating from the surrounding stainless steel matrix. The fact that the brown stain grew beneath the resin indicates that excess iron (or steel) was present on the stainless steel surface before the resin application. Examination of the precision bonding procedure indicated that in it the stainless steel part was held in place by a soft steel screw that smeared the surface of the stainless steel part. This soft steel smear entrapped moisture and slowly oxidized.

Recommendation. A teflon holding screw replaced the soft steel screw, which solved the problem.

CASE STUDY 6.7: Mineral Deposits

Problem. A voice coil machine exhibited a white coating during routine inspection.

Hypothesis. Corrosion of the part was suspected.

Analysis. Both copper and steel parts of the machine exhibited this white coating. Figure 6.6 is a SEM micrograph showing the crust-like formation. EDXS analysis showed the presence of calcium with small amounts of aluminum, silicon, and iron, with traces of sodium and chlorine. Powder was scraped from the surface and studied with X-ray diffraction (XRD), which showed peaks representing calcium-aluminum-silicate hydrate.

Discussion. Because both copper and steel exhibited the same residue, corrosion was unlikely as the mechanism of contamination. Moreover, iron oxide has a characteristic reddish brown rust color. When the crust was scraped off, the surface underneath was clean and smooth, further ruling out corrosion. The major source of calcium salts is hard water with dissolved minerals. Investigation of the manufacturing process revealed that the parts were cleaned in

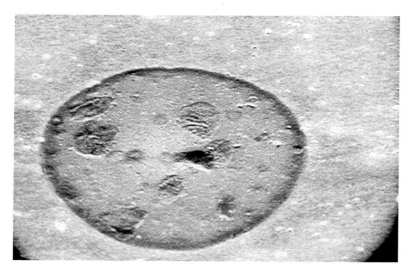

Fig. 6.7 Human handling contamination (e.g., spittle) baked onto the substrate. EDXS analysis showed the presence of chlorine, potassium, and sulfur. SEM, 200×.

hard instead of deionized water, from which minerals were deposited on them.

Solution. When deionized water replaced hard water for cleaning, the problem was eliminated.

CASE STUDY 6.8: Human Contamination

Problem. On an oxide disk manufacturing line, numerous "hits" or obstructions occurred when completed disks were fly tested with a head. This is usually caused by particulate contamination. To determine its source, 20 dummy aluminum disks were sent through the manufacturing line and studied for the presence of contamination.

Analysis. Contamination spots were first studied in an optical microscope and then by SEM analysis. They could be categorized into two groups. One group appeared to have dripped into a round drop and hardened (see Fig. 6.7). The EDXS analysis showed the presence of chlorine, potassium, and a small amount of sulfur. The spot adhered strongly to the substrate and could not be wiped off with a cotton swab. The second group (see Fig. 6.8) appeared to have

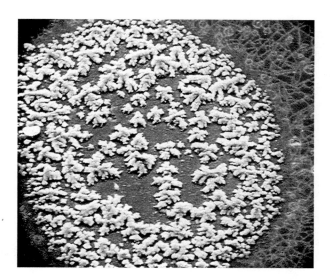

Fig. 6.8 Human handling contamination (e.g., cosmetic material) that fell on the disk during the final stages of manufacturing. EDXS examination showed the presence of chlorine, sodium, potassium, and calcium. SEM, 3900x.

Fig. 6.9 White powder on aluminum (the chemical reaction product) and bubbling of the coating, indicating a chemical reaction with aluminum.

Fig. 6.10 Flaking of oxide medium coating caused by the chemical reaction of aluminum with nitric acid. SEM, 240×.

dried on the surface. The EDXS analysis revealed many particles consisting of chlorine, potassium, sodium, and calcium.

Discussion. The first group of spots consisted of human contamination, such as spittle or perspiration, baked onto the substrate. The oxide medium disk is baked at approximately 180 °C (355 °F) for several hours, to cure the binder resin (see Appendices 2.2 and 2.3 for details of oxide media manufacturing). Contamination occurred before the baking cycle, and because it was baked on the disk, it adhered strongly to the surface.

The second group of contamination spots were human cosmetics, such as antiperspirants, deposited after the lubrication process and not disturbed afterwards. They were essentially human salts and

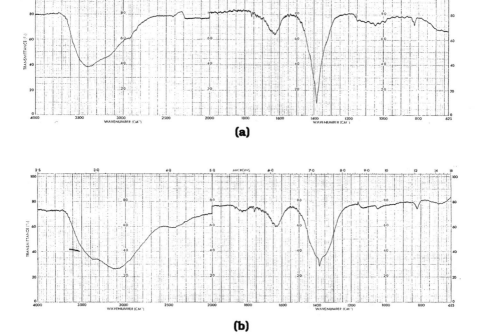

Fig. 6.11 FTIR spectra from KBr pellet of (a) white powder on aluminum and (b) aluminum nitrate, showing an excellent match between the two spectra.

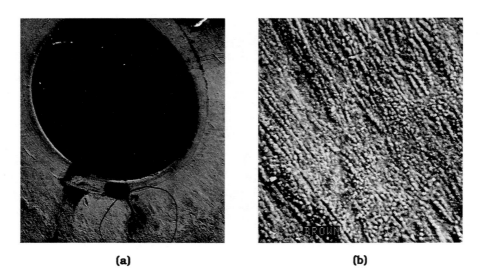

(a)   (b)

Fig. 6.12 Brownish area on Sn-Pb solder on a printed circuit board. (a) Low magnification. (b) High magnification.

cosmetics that remained after the liquid evaporated and could be partially wiped off with a cotton swab.

Solution. When good clean-room practices, e.g., use of gloves, masks, etc., were followed stringently, the contamination problem was eliminated.

CASE STUDY 6.9: Nitric Acid Attack on Aluminum

Problem. A disk drive crashed and was opened for investigation. The oxide medium disk exhibited bubbling of the coating and white powder on the aluminum substrate.

Analysis. Figure 6.9 is a SEM micrograph of the white powder on the aluminum and bubbled resin coating. Figure 6.10 shows lifting of the coating from aluminum substrate corrosion. The FTIR spectrum of a KBr pellet made of the white powder appears in Fig. 6.11(a), and Fig. 6.11(b) is the FTIR spectrum of aluminum nitrate.

Discussion. The micrographs showed that a liquid seeped into the disk drive along the spindle and flowed over the disk while it was spinning. The bubbling of the oxide coating indicates that a chemical reaction with aluminum occurred. One of the common chemicals

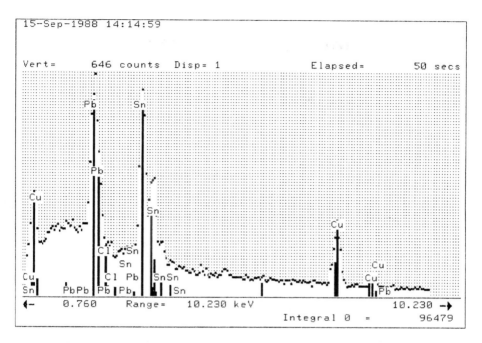

Fig. 6.13 EDXS spectrum of brown area showing copper beneath the solder, which is probably an organic substance. There was no corrosion.

that corrodes aluminum in this manner is nitric acid. The two FTIR spectra shown in Fig. 6.11 match very well, indicating that the white powder was aluminum nitrate.

Analysis confirmed that nitric acid removed the oxide coating and the head-disk crash. The fact that the attack was confined to the inner radius of the disk showed the corrodent came from the spindle. Investigation revealed that nitric acid from a spill near the disk drive acid traveled along the spindle into the drive.

Solution. There was nothing wrong with the drive itself. The nitric acid spill caused the crash.

CASE STUDY 6.10: Discoloration Versus Corrosion

Problem. On a printed circuit (PC) board, brown discoloration was noticed on top of the eutectic Sn-Pb solder.

Hypothesis. Corrosion or tarnishing of the solder was suspected.

Analysis. Figure 6.12(a) is a low-magnification micrograph of the brown spot on the eutectic Sn-Pb solder, and Fig. 6.12(b) is a

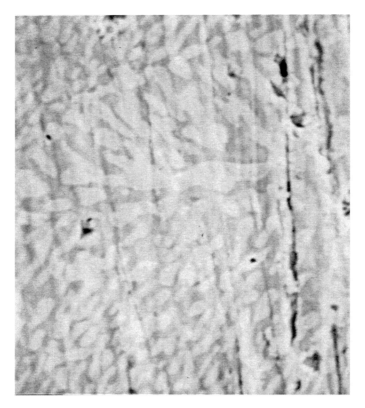

Fig. 6.14 Shiny area of solder with the Sn-Pb eutectic composition.

high-magnification micrograph of the same area. Figure 6.13 provides the EDXS spectrum from the discolored area, which reveals the presence of copper beneath the solder. A normal, shiny area of solder with the typical eutectic structure is shown in Fig. 6.14, and Fig. 6.15 is the EDXS spectrum of that area revealing the presence of only tin and lead.

Discussion. Under identical EDXS conditions, the fact that the presence of copper shows in the spectrum in Fig. 6.13, and not in Fig. 6.15, shows the solder was much thinner in the brown-colored area. The presence of copper beneath the thin solder film probably gave a brown tinge to that area. Because there was no indication of the presence of chlorine or sulfur, corrosion could be eliminated as

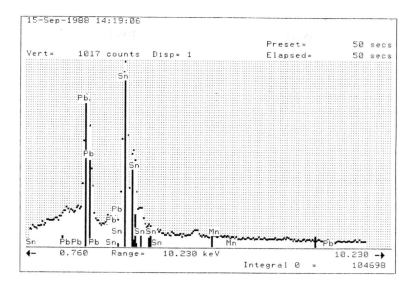

Fig. 6.15 EDXS spectrum of the eutectic area. Normal thickness of solder reveals the presence of copper in EDXS analysis.

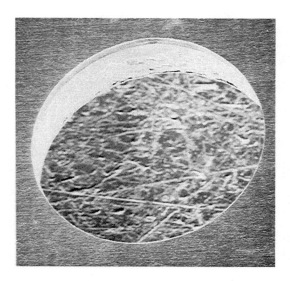

Fig. 6.16 Hole in a type 302 stainless steel part showing possible contamination on the inner top of the hole.

a possible source of contamination. Apparently, a hard object rubbed against the solder and deposited an organic material. The mechanical abrasion removed a portion of the solder, exposing the copper underneath during EDXS analysis.

Solution. Tarnishing does not always indicate a corrosion mechanism. Mechanical abrasion may cause discoloration. Although there was nothing wrong with the PC board, it should have been handled more carefully.

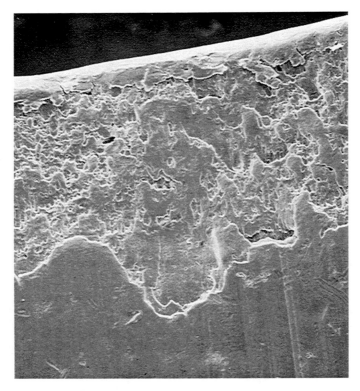

Fig. 6.17 Same area exhibiting false impression of possible contamination. The imperfection was caused by the hole-punching process.

CASE STUDY 6.11: Improving Cleaning Techniques

Problem. A type 302 stainless steel part exhibited possible contamination on the inside of a hole. Several cleaning techniques were tried unsuccessfully.

Hypothesis. A hard baked organic substance was the suspected source of contamination in the hole.

Analysis. Figure 6.16 is a micrograph of an imperfection found inside the hole. Mechanical abrasion, organic cleaning, and nitric acid cleaning were unsuccessful at removing the contamination. Closer examination of the area (Fig. 6.17) showed no contamination. It was an imperfection caused by the hole-punching procedure.

Discussion. The punch was made of hardened steel. Because nitric acid did not dissolve the possible contamination, it was found to be a defect in the stainless steel and was not transferred to the steel during processing. The rough edges of the punch probably caused the defect.

Solution. The discoloration was not a serious defect. The hole-punching process may need to be optimized. Many cleaning techniques were tried based on the assumption that contamination was

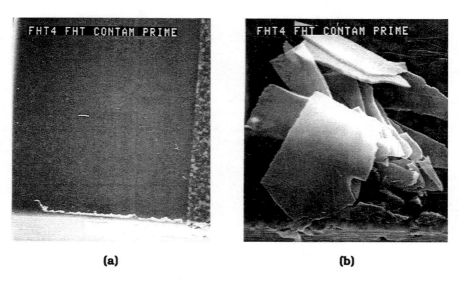

(a)   (b)

Fig. 6.18 Makeup material that is a source of handling contamination. (a) Low magnification. (b) High magnification.

Fig. 6.19 Cotton fibers on a cotton swab. They are porous and weak and generate loose-fiber contamination.

present, but they were unsuccessful. Consequently, the composition and source of contamination should be known before implementing a cleaning procedure.

CASE STUDY 6.12: Handling Contamination—Cosmetics

Problem. During fly testing of thin-film heads over disks, some contamination collected on the head.

Hypothesis. It was assumed that the disk was generating debris.

Analysis. Figures 6.18(a) and (b) are SEM micrographs of the contamination at different magnifications. The EDXS analysis indicated the presence of chlorine, potassium, and sodium.

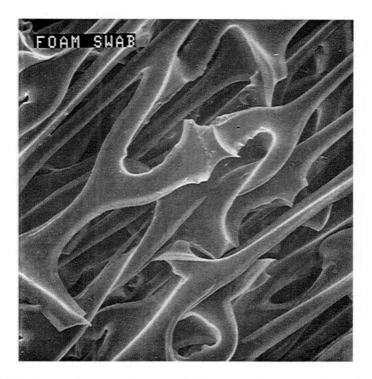

Fig. 6.20 Foam fibers on a foam swab. They are stronger and are more uniform in size, and while they do not absorb as well as cotton, they have improved contamination performance.

Discussion. Because of the design of the head rails, contamination generally accumulates at the outer edges. From EDXS analysis, it is clear that the contamination was caused by human handling. The morphology indicates that cosmetic material fell on the disk and accumulated on the head.

Solution. Clean-room personnel should not wear extensive cosmetics that could become airborne and enter the clean-room environment.

CASE STUDY 6.13: Are Cotton Swabs Safe?

Problem. Organic fiber contamination was found in the clean-room manufacturing area. Should cotton swabs be allowed in the clean room?

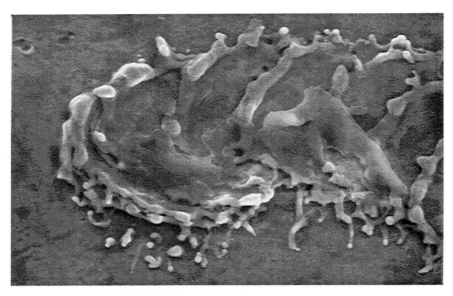

Fig. 6.21 Laser-scribed stainless steel that could generate loose metal particles.

Analysis. Figure 6.19 is an SEM micrograph of cotton on a swab, and Fig. 6.20 is that of foam on a foam swab.

Discussion. The cotton fibers have a wide size and shape distribution. Loose fiber pieces are also visible. The foam fibers are uniform in size and are larger in diameter than the cotton fibers. The porosity of the cotton fibers makes them better absorbers of solvents and as such are better cleaners. However, they generate many loose fibers. Foam fibers are much better than cotton from the standpoint of contamination.

Solution. Replace cotton swabs used in the clean-room environment with foam swabs.

CASE STUDY 6.14: Scribing of Stainless Steel

Problem. Did scribing of stainless steel parts for process monitoring cause any contamination problems?

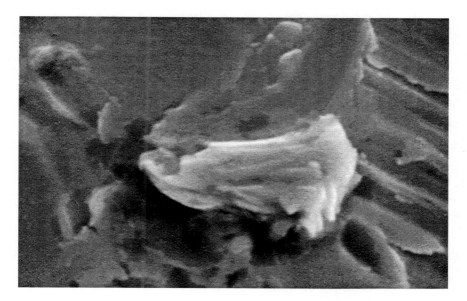

Fig. 6.22 Scribing of stainless steel by mechanical abrasion could also generate metal contamination.

Analysis. Figures 6.21 and 6.22 are SEM micrographs of stainless steel parts that were scribed with a laser and an abrading machine tool.

Discussion. The laser melted the metal, which beaded up as it cooled. The laser scribing parameters need to be optimized so that such beading does not occur. The abrading tool plastically deforms and fractures the metal. If not cleaned properly, the metal could be coated with loose metal debris.

Solution. The scribing process should be optimized based on the generation of contamination.

References

1. B. Bhushan, *Tribology and Mechanics of Magnetic Storage Devices,* Springer-Verlag, 1990

2. W. Prater, W. Jones, G. Stone, and J. McDowell, Preventing Contamination in Magnetic Disk Drives Through the Use of Wear-Resistant Coatings, *Microcontamination,* April 1989, p 31

3. P. Viswanadham, Contamination Control in Disk-Drive Manufacturing—A Quality and Reliability Perspective, *Microcontamination,* April 1987, p 41

4. A. Arora, Surface Contamination Measurement and Control by Nondestructive Techniques, *J. Environ. Sci.,* Nov-Dec 1985, p 30-32

Appendix 6.1

Head/Media Interface: The Heart of a Disk Drive*

Important functions are performed by the electronics and mechanics of a disk drive, and the interface between the read/write head and the disk itself is the heart of the drive. When the head/disk combination performs its intended function properly, the drive can meet or exceed its specifications. The drive can be counted on to handle large volumes of information and to deliver data quickly, on demand. However, a head/disk interface malfunction, due either to inappropriate design or improper working environment, can lead to a head crash. The head/disk interface is the main source of disk drive frailty. Consequently, the most productive way to enhance the reliability of a drive is to optimize this interface.

Drive Technology

Popular contemporary hard disk drives use magnetic-layered platters made with one of two technologies; particulate oxide or metal. For the former, an aluminum substrate is coated with a paint containing polymer resins and particles of iron oxide with a small percentage of alumina. A fluorocarbon lubricant, 5 nm thick, is added. On a metallic disk, a magnetic thin film is either plated or is vacuum sputtered onto the substrate. A metallic disk also has a 30-nm carbon isolation layer and may have a 3-nm fluorocarbon lubricant layer.

---

* P.B. Narayan and A.S. Brar, *Computer Technology Review*, Spring 1989, p 26. Reprinted with permission.

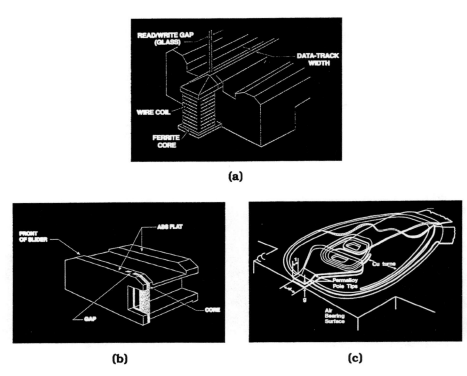

(a)

(b)   **(c)**

Fig. 1 The three basic types of read/write heads. (a) Monolithic, (b) Composite, (c) Thin-film.

The three basic types of read/write heads are monolithic, composite, and thin-film (Fig. 1). Monolithic heads are made of ferrite-type magnetic ceramics. In a composite head, a ferrite core is encased in a ceramic head pad material. In a thin-film head, the core and the coil turns are sputtered on a ceramic substrate. An advanced version of another head type, magnetoresistive, is made in the same way as thin-film heads. However, the signal is picked up by the magnetoresistive effect instead of magnetic induction. The most popular head pad ceramic materials are calcium titanate, Alsimag, and barium titanate.

Catamaran-type heads are widely used at present. Research is being done on negative pressure air bearing heads, which may be useful to obtain better reliability at lower flying heights.

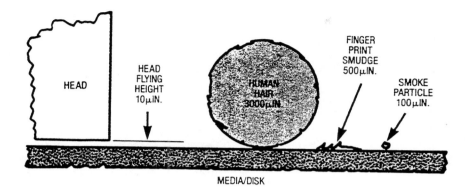

Fig. 2 Hydrocarbon or particulate contamination at the head/disk interface can lead to friction buildup and distortion of the head arm assembly. This figure illustrates the relationship of the head flying height to various

Aerodynamics

In modern Winchester technology disk drives, the read/write head rests on the disk when the drive is turned off. When power is applied, the disk starts to rotate and an aerodynamic lifting force is applied to the head. The head becomes fully airborne when the disk reaches a rotational speed of about 1000 rpm, and flies on an air bearing at a specified height (about 250 nm) when the disk reaches its designed speed, usually 3600 rpm. When the drive is turned off, the head lands on the disk.

Enormous stresses are created at the head/disk interface when the head is taking off or landing and the head and disk are in physical contact. Head destabilization can also be caused by accidental contact of the disk and the flying head during operation. In addition, hydrocarbon or particulate contamination at the interface can lead to friction buildup and distortion of the head arm assembly (Fig. 2). Any of these eventualities can lead to a catastrophic head crash.

Materials Selection

Optimum materials selection, design, and processing are an integral part of a successful interface. Among the goals are low friction, low wear, and reliable lubrication. These are achieved (after proper design and concern for materials compatibility) in a large part by proper surface preparation. The upper atomic layers of the mating parts, after all, determine the tribological properties of the interface, e.g., how much friction and wear the interaction of the two surfaces will cause.

Too little lubricant on the disk causes unacceptable wear, and too much lubricant leads to high static friction, also known as stiction. As the disk rotates, inertial forces tend to make the lubricant fly off the disk surface. Consequently, the lubricant has to be constantly replenished. On particulate iron oxide disks, the oxide media has many pores that act as lubricant reservoirs. The lubricant anchoring mechanism is a little more complex on metallic disks.

The disk is often textured with concentric grooves to optimize tribological and magnetic properties. Texturing provides a lubricant reservoir and reduces the area of contact of the disk with the head, thus reducing friction. The water absorption characteristics of the disk surface are also taken into account, because absorption of moisture by the carbon overcoat and the lubricant significantly alters the tribology of the interface.

Pad Materials

In monolithic heads, the core and head pad materials are the same. In the other head types, the head pad material has a substantially larger area exposed to the disk than the core magnetic material that has the greatest effect on head/disk tribology.

Material toughness is very important. The head pad material should not chip or crack during processing and must be tough enough to absorb the stresses of takeoff and landing without fracturing. Chipping and cracking generate dangerous particulate contamination.

The addition of suitable tribo-materials reduces the friction coefficient of the interface, which is valuable for design as well as reliability considerations. The tribo-materials cover weak areas on the surface, too.

The head pad material should have a low surface energy, because high-energy surfaces tend to attract contamination from the surrounding atmosphere. The material should also have a low affinity toward the disk constituents, because a high affinity leads to delamination of the surface layers, increasing friction, and head buildup.

The head pad material is lapped and polished to obtain a smooth surface finish, one with a surface roughness of 20 nm. Because the disk is only in contact with the top layers of the head pad material, damage caused by lapping and polishing must be minimized. Lapping can cause deformation of the crystal structure of the material and nonuniformity of the material hardness across the surface. Consequently, lapping damage should be minimized by optimizing process variables such as lapping load and duration, lapping compound composition, and grit size. Materials properties such as microtoughness, microhardness, surface energy, and thin-layer stresses have to be studied and analyzed thoroughly to understand how the interface truly functions.

Debris

The hard alumina particles in the media of oxide disks can remove some of the debris that collects on the heads. However, too much debris can still lead to a malfunction of the interface.

Debris can come from the media itself, or from other components inside the drive. Many components are made with aluminum alloys and undergo precise machining and other processing operations. These components must be protected from atmospheric corrosion and should not be allowed to generate particulate or film-like contamination. They are coated via one or more of the many surface coating techniques such as chromate conversion, anodizing, organic coating, ion vapor deposited aluminum, or electroless nickel. The most suitable coating depends on the application.

There are many plastics, gaskets, adhesives, and other polymer materials used in a drive. They should not outgas, decompose, or generate contamination in the drive. All the polymeric and metallic components must be thoroughly tested for outgassing, cleanliness, and contamination generation.

Appendix 6.2

Analyzing Spittle Mark Contamination on Computer Disks*

In peripherals manufacturing such as the computer disk drive industries, the trend is toward ever-increasing storage capacities as the physical size of the drives continues to decrease. Consequently, media parameters such as the magnetic medium thickness, the flying height of the read/write head above the disk, and the head gap length also must be reduced.

Generally speaking, the dimensional tolerances in the data storage industry are even more critical than in semiconductor technology. In a typical high-density disk drive, the flying height of the head above the disk is less than 0.25 µm, and the magnetic layer is 0.5 µm thick for the particulate oxide medium and 0.05 µm for the thin-film metallic medium.

Human Handling

There is a strong drive in the peripherals industry to automate manufacturing to decrease the amount of handling, with robots taking over many manufacturing functions. However, various intricate processing steps still require human handling. Until recently, it was not fully appreciated by many in the industry that even careful human handling could introduce contamination. Thus, clean-room practices, such as wearing face masks, gloves, and other protective garments, have not been very popular and were not well appreciated.

---

* A.S. Brar and P.B. Narayan, *Microcontamination*, Sept 1988, p 67. Reprinted with permission.

A few years ago, it was not uncommon to find workers moving around without face masks in a clean room.

We need to convince all manufacturing personnel of the need for stringent adherence to good clean-room practices. One problem has been that the nontechnical worker was not provided with a clear demonstration that handling introduces some of the worst microcontamination possible. The most prominent sources of this contamination are spit marks and fingerprints.

Spit Mark Contamination

A detailed study follows regarding spit mark contamination found on hard disks used in memory disk drives. An unusually large number of failures due to catastrophic head crashes from one lot of particulate iron oxide disks prompted close study of the disks and heads. Evidence of microcontamination, such as spit marks and fingerprints, caused by human handling was found.

A spit mark is barely visible to the naked eye, yet is clearly visible in an optical microscope due to the interference colors produced by the remnant layers of spit (Fig. 1). The specimen area is highlighted with a marker and analyzed in a scanning electron microscope (SEM). Because of the poorer contrast in a SEM, it is always easier to locate the mark in an optical microscope and highlight the area before the SEM examination.

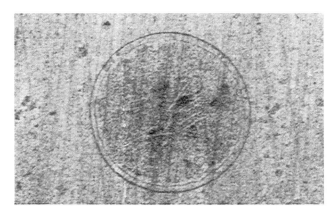

Fig. 1 Optical micrograph of spit mark is clearly visible because of the interference colors produced by very thin films in light microscopy.

Fig. 2 SEM micrograph of spit mark showing the dendritic structure of
the crystals solidifying out of the spit liquid.

Microscopic Analysis

SEM and EDXS Analysis. Figure 2 is a SEM micrograph of a spit
mark that is more than 100 µm wide. There appears to be a dark
organic material (about 40 µm wide) inside the mark with salt
solution surrounding it. One can see the dendritic structure of
crystals solidified out of the evaporating spit liquid.

Energy-dispersive X-ray spectroscopy (EDXS) analysis of the spit
mark indicates the presence of aluminum, iron, chlorine, and potassium (Fig. 3). The aluminum and chromium signals come from the
chromate-coated aluminum alloy substrate of the rigid disk, and the
iron signal originates from the iron oxide in the magnetic medium.
The gold signal is from the gold coating put on the specimen to
prevent charging. Thus, with EDXS, only the bare presence of
chlorine and potassium can be detected in the spit marks, indicating
that EDXS has only limited use in chemically analyzing very thin
layers.

Even though the electron beam size in EDXS is several hundred
angstroms, the X-ray signal originates from an area a few microns
deep. Because EDXS is not surface sensitive enough for this purpose, Auger electron spectroscopy (AES) was used.

AES Analysis. In AES, a beam of electrons excites auger transitions in the specimen, and the analysis depth of Auger electrons is

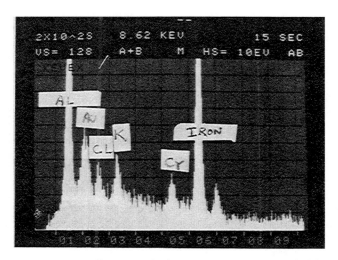

Fig. 3 EDXS analysis of spit mark showing potassium and chlorine (in addition to aluminum and iron coming from the hard disk constituents). Chromium comes from the chromate coating on aluminum. Gold is from the gold coating deposited to prevent charging.

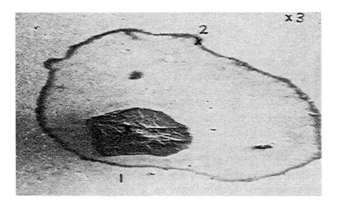

Fig. 4 SEM micrograph of the spit mark that was analyzed at three different points by AES.

less than 10 nm. Thus, AES provides the chemical compositions of the first few atomic and molecular layers.

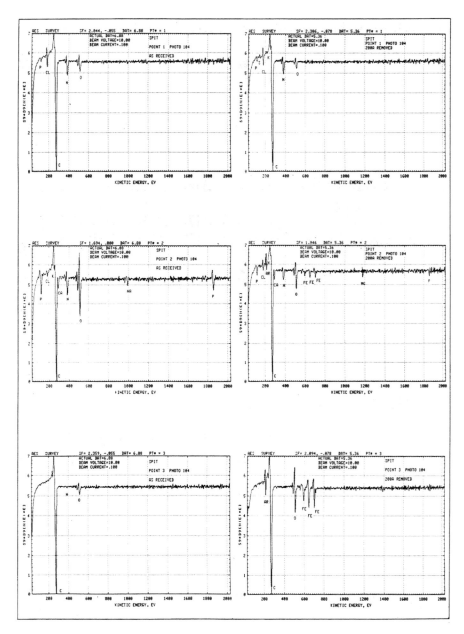

Fig. 5 AES spectra from the three locations shown in Fig. 4 in the as-received condition and after a depth of 20 nm (see Table 1 for a summary of surface chemical compositions at these locations).

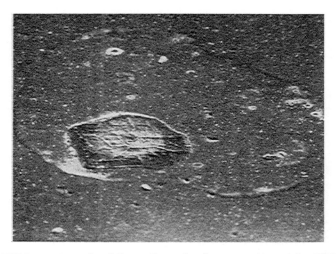

Fig. 6 SEM micrograph of the spit mark after a sputter etch of 20 nm showing the virtual disappearance of the mark except the central organic agglomerate (probably a few thousand angstroms thick).

Figure 4 is a SEM micrograph of a spit mark. The AES analysis was performed at three points: on the central organic core material, at the outer edge of the dried solution mark, and on the substrate away from the spit mark. Because any material exposed to the atmosphere has a few monolayers (up to 5 nm) of absorbed hydrocarbons and other carbon compounds, a couple of hundred angstroms of material were routinely removed by argon ion sputter etching. AES analysis was repeated to provide the surface composition of the specimen.

The chemical compositions at the three points before and after sputter etch are shown in Fig. 5 and summarized in Table 1. At the third point, carbon, some oxygen, and a trace of nitrogen are visible before the etch. After the etch, carbon, oxygen, and iron are visible, showing the iron oxide and the binder in the magnetic medium.

At the second point before the etch, phosphorus, carbon, potassium, calcium, and sodium are present, in addition to the atmospheric constituents of carbon, oxygen, and nitrogen. After the etch, iron can be clearly seen, indicating the layer is about 20 to 30 nm

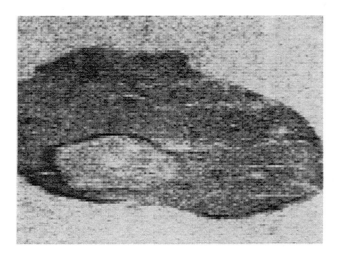

Fig. 7 Auger carbon mapping of the area in Fig. 4, showing the dominant presence of carbon at the central agglomerate.

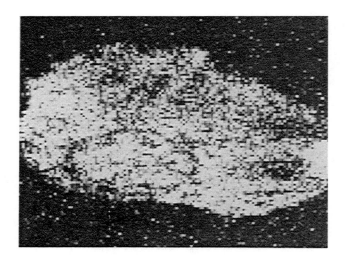

Fig. 8 Auger nitrogen mapping of the same area showing the general presence of nitrogen in the area (with higher concentrations at the agglomerate).

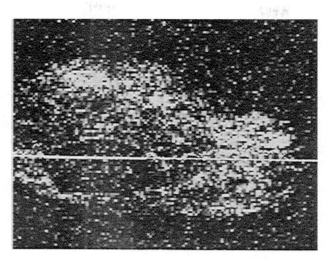

Fig. 9 Auger sodium mapping showing its concentration at the outer edges.

thick. The concentration of water-soluble salts such as potassium chloride and sodium chloride is visible at the outer edge of the mark.

At the first point, a strong peak of nitrogen is present even after etch, indicating that the organic material is mostly carbon, nitrogen, and oxygen. Figure 6 shows the spit mark after the 20-nm etch, showing the virtual disappearance of the spit mark except the central organic agglomerate, which appears to be a few thousand angstroms thick.

The above analysis indicates that, as the spit mark dries, the proteins, enzymes, and tissue material stay as an agglomerate that is a few thousand angstroms thick at the center. The liquid, rich in metallic chloride salts, spreads to a thin sheet of about 20 to 40 nm thickness, due to surface tension. Thus, the metallic chlorides are enriched at the outer edges of the spit mark and become crystallized in a dendritic structure.

SAM Analysis. The above conclusions can be summarized by using scanning auger microscopy (SAM), which essentially involves Auger mapping for various elements on the specimen surface. The spit mark (Fig. 4) is mapped for carbon (Fig. 7), nitrogen (Fig. 8), and sodium (Fig. 9). Thus the SAM analysis indicates that sodium is concentrated at the outer edges, and carbon and nitrogen are concentrated at the central organic agglomerate. Figure 10 clearly dem-

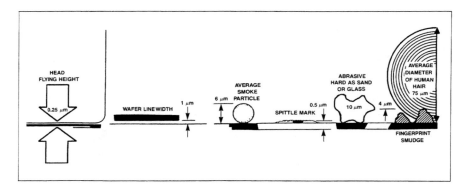

Fig. 10 Approximate sizes of common contaminants compared with the
critical dimensions in disk drives and semiconductor devices.

onstrates the approximate sizes of common contaminants compared
with the dimensions obtained in disk drives and on silicon chips.

Conclusion

Spit marks are about 100 μm wide, consisting of a central organic
agglomerate of proteins, enzymes, and tissue material (about 40 μm
wide and several tenths of a micron thick) and a 20 nm thick layer
of such metallic salts as sodium chloride and potassium chloride.
Contamination several tenths of a micron thick can easily cause
head crash in a disk drive, because the flying height is less than 0.3
μm. Although the corrosion properties of the chlorine species introduced by the spit marks have not been considered in the present
analysis, it can be a serious problem because the thin-film, metallic,
magnetic medium is about 0.05 μm thick.

The dimensions of spit marks indicate that they can be detrimental to the performance of components in semiconductors, magnetic
recording, and other devices. Human handling can introduce microcontamination into the systems, but stringent adherence to sound
clean-room practices eliminates the problem.

Chapter 7

Head-Disk Interface

The head-disk interface is the most critical part of the disk drive, especially from the standpoint of reliability. Often minor defects in disks and heads are not found until the disk drive is fully assembled. It is much more cost-effective to detect defects as early in the manufacturing process as possible. Thus, when a head-disk failure is analyzed, it is important to determine the earliest point in the manufacturing process at which the defect existed or originated. Several micrographs of heads and disks are presented below, and the methodology of determining the origin of the respective defects is discussed.

Figure 7.1 shows a surface pore defect on a hard disk with particulate iron oxide magnetic medium. (See Appendix 7.1 for details of manufacturing and testing a hard disk with oxide medium.) The area surrounding the defect appears to be undisturbed, indicating that the defect existed before the substrate was coated with the magnetic oxide paint. Aluminum-magnesium alloys, of which the substrate was made, generally have intermetallic compounds containing aluminum, iron, silicon, and manganese. During substrate lapping and polishing, these intermetallics can be pulled out of the surface, leaving a pore where they were removed. The magnetic coating thickness is nonuniform at the defect site, which causes deterioration of the read signal. Also, the surface pore can trap moisture, etching solution, or cleaning solution, which can cause corrosion.

Figure 7.2 shows another defect on an oxide disk. The site was covered with a thin oxide coating. Energy-dispersive X-ray spectroscopy (EDXS) analysis indicated the presence of aluminum, iron,

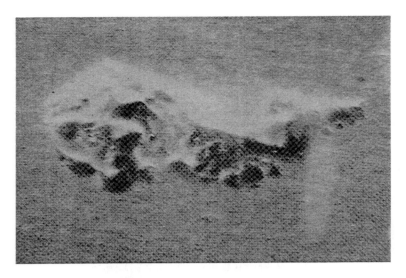

Fig. 7.1 Morphological inhomogeneity on aluminum disk, which caused defective oxide media coating. SEM, 11,600×.

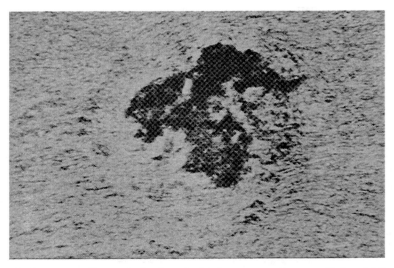

Fig. 7.2 Intermetallic particle chemical inhomogeneity on a disk, which caused defective media coating. SEM, 11,600×.

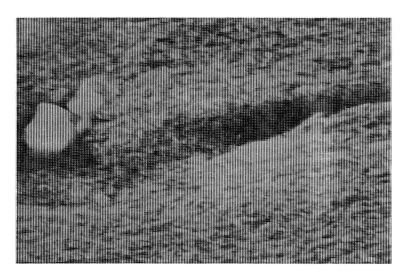

Fig. 7.3 Alumina agglomerate produced in the manufacture of oxide media caused scratches in the direction of disk rotation during buffing. SEM, 17,400×.

silicon, and manganese. Because the coating in the surrounding area was undisturbed, it is evident that this defect also existed before the oxide coating was applied. The EDXS composition analysis revealed the presence of an intermetallic compound on the aluminum surface. Both the chromate conversion undercoat and the oxide coating did not adhere well to this compound. As a result, very little coating remained on the disk, thus drastically reducing the read signal amplitude. These sites can also act as fluid entrapments. If these compounds are pushed into the aluminum substrate, they may pass through the burnishing operation and appear as "hits," or projections out of the surface later on because of relaxation of the aluminum matrix over a period of time, e.g., several months.

Figure 7.3 shows a scratch defect on an oxide disk. An alumina agglomerate formed at the head of the scratch during preparation of the oxide paint. Figure 7.4 is a transmission electron microscope (TEM) micrograph of an agglomerate. These scratches are fine and circumferential, i.e., in the polishing direction. During polishing of the magnetic layer after curing, alumina particles were loosened from the surface and caused the scratch. The fact that the head did not exhibit the presence of any noticeable media debris and alumina indicated that the disk was scratched before the head and disk were

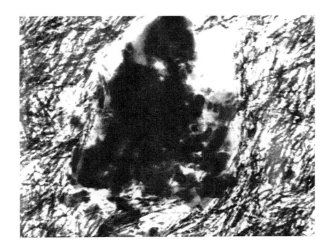

Fig. 7.4 Alumina agglomerate in the oxide media. TEM, 12,000x.

assembled. Such scratches cause signal loss and loose contamination.

Figure 7.5 shows alumina particles that were present on the disk before it was assembled. Because these particles are not embedded in the coating, they did not originate from alumina agglomerates formed during media manufacturing. The coating is polished with tape impregnated with alumina particles, and the tape disintegrated, causing the transfer of alumina particles to the disk.

Figure 7.6 shows a scratch that is relatively deep in the aluminum substrate on an oxide disk. Because it is random in direction, polishing was ruled out as the source of damage. The scratch was caused by handling damage, in which a sharp edge, such as a tool, accidently touched and scratched the disk.

CASE STUDY 7.1: External Contamination

Problem. In one lot of drives, defects 50 µm (1970 µin.) in diameter appeared suddenly on an oxide disk when the drives were running.

Analysis. Figures 7.7 and 7.8 illustrate typical disk defects. At most defect sites, only aluminum and iron disk constituents were

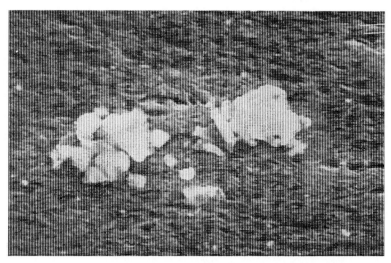

Fig. 7.5 Loose alumina particles from the buffing tape that adhered to the media surface. SEM, 17,400×.

Fig. 7.6 Deep, random-direction scratch caused by mechanical damage from improper handling. SEM, 11,600×.

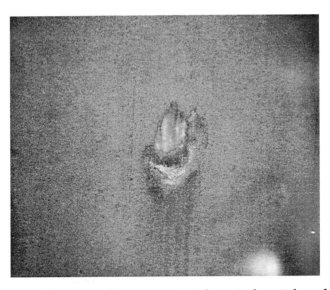

Fig. 7.7 Disk defect, caused by a stray stainless steel particle, enlarged when the head flew over it. 250×.

found by EDXS analysis. However, a few defects exhibited bright particles that contained iron, chromium, and nickel (Fig. 7.9). The thin-film head exhibited significant disk media debris (Fig. 7.10), in which a few particles showed the presence of iron, chromium, and nickel, per the EDXS spectrum in Fig. 7.11. The indications of aluminum and titanium signals were coming from material removed from the the head slider, made of a composite ceramic of alumina and titanium carbide. During the manufacture of thin-film heads, the magnetic yoke is deposited by thin-film deposition techniques available in the semiconductor industry on a ceramic material called head slider material. At the head-disk interface, from the point of view of area, most of the head material that interacts with the disk is slider material. The presence of iron, chromium, and nickel indicated a type 300 series stainless steel particle.

Solution. The defect contained a gouge that increased in depth in the direction of the disk rotation, indicating the presence of a stray particle, harder than aluminum, that hit the rotating disk. From the end of the gouge, scratching of the magnetic coating was evident, but no significant gouging of the aluminum was apparent. When the stray particle hit the disk, displaced aluminum substrate material

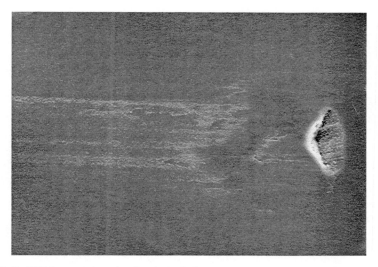

Fig. 7.8 SEM micrograph of a disk defect made by an impinging hard particle of stainless steel.

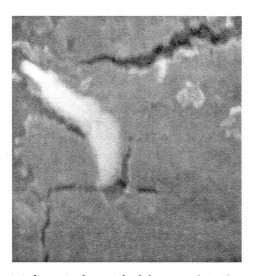

Fig. 7.9 Bright stainless steel particle debris stuck in the media.

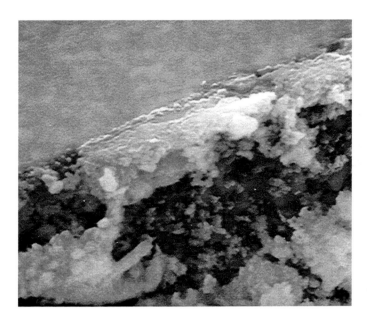

Fig. 7.10 Debris on the head, primarily oxide media scraped off the disk, with stainless steel particles from head-disk contact.

accumulated at the edge of the gouge. While the head had not encountered this defect up to this point, when it flies over the defect site it will remove the buildup of aluminum, scratching the media coating, which is much softer than aluminum, in the direction of the head flight. Next to aluminum, stainless steel is the most common material used in disk drives and is much harder than aluminum. A stray stainless steel particle hit the disk, causing the beginning of the defect. The fine particles of stainless steel that accumulated on the head and in the disk scratch confirmed the presence of stainless steel particle contamination in the drive environment.

Recommendation. All potential sources of stainless steel contamination in the drive should be removed.

CASE STUDY 7.2: Defect Growth on Plated Media

Problem. In a drive with plated magnetic media and a thin-film head, the head suddenly crashed.

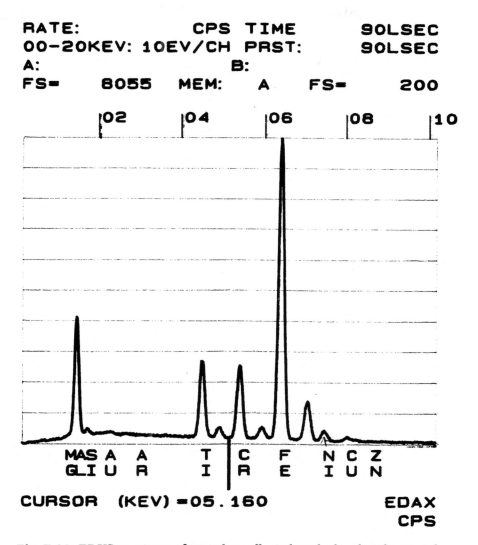

RATE: CPS TIME   90LSEC
00-20KEV: 10EV/CH PRST: 90LSEC
A: B:
FS=   8055   MEM: A   FS=   200

CURSOR   (KEV) =05.160   EDAX
                                          CPS

Fig. 7.11 EDXS spectrum of particles collected on the head, indicating the presence of iron, chromium, and nickel (type 300 series stainless steel), in addition to aluminum and titanium from the head slider material.

Analysis. On the disk, the defect was approximately 100 μm (3940 μin.) in diameter (Fig. 7.12). The EDXS study revealed the presence of silicon and calcium, indicating calcium silicate dust contamination. At what stage of manufacturing did the dust particle

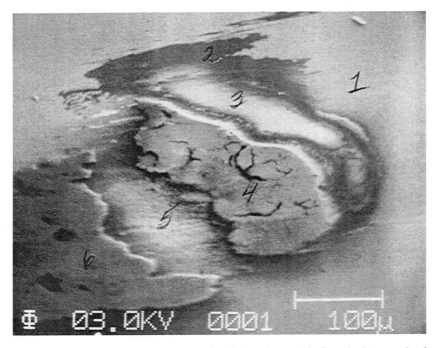

Fig. 7.12 Dust contamination on the electroless nickel underlayer, which caused water entrapment and corrosion, leading to a head-disk crash.

originate? The EDXS examination showed the presence of silicon, calcium, cobalt, nickel, and phosphorus. The plated magnetic medium was a Co-Ni-P alloy, 30 nm (1.18 μin.) thick plated over 5 μm (197 μin.) thick electroless nickel, which contains nickel and phosphorus. Auger electron spectroscopy (AES) indicated that, outside the defect area, carbon (carbon overcoat) was present. On the contamination, cobalt, nickel, and phosphorus (plated media) were present, and on the area beneath the contamination, nickel and phosphorus (electroless nickel underlayer) were present.

Solution. The AES study revealed that the dust contamination was plated over by the cobalt alloy, which indicates that the contamination occurred during electroless nickel polishing. The contamination retained plating solution and water. The cobalt alloy corroded slowly, and the area swelled, protruding into the path of the head

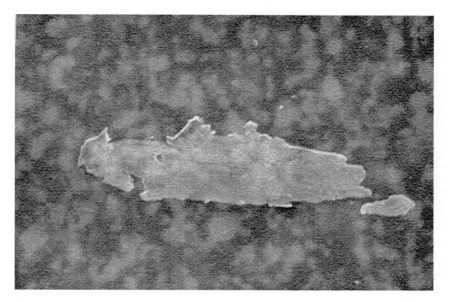

Fig. 7.13 Soft solder (Sn-Pb) accumulation at the head-disk interface that was smeared over the head surface.

flight. The flight of the head was disturbed violently, forcing it to crash into the disk.

Recommendation. The electroless nickel polishing step was purged of all dust contamination.

CASE STUDY 7.3: Accumulation of Soft Material at the Head-Disk Interface

Problem. The head accumulated contamination on the surface of the air bearing after testing.

Analysis. Figure 7.13 shows that the contamination was smeared over the air bearing surface. The EDXS analysis indicated the presence of tin and lead.

Solution. The presence of tin and lead indicated that solder material, which is softer than either the glass disk (used for fly testing) or the head material, had been introduced into the drive environment. Solder has comparable hardness to the disk, and if it accumulated at the head-disk interface, it would have damaged the disk. Because no damage was noticed on the disk, it was likely the

 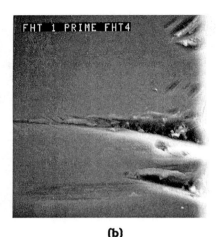

(a)   (b)

Fig. 7.14 Hard dust particle at the head-glass disk interface, which scratched the alumina. The gouge is deepest at the edge of the alumina, the point of lowest flying height. (b) Higher magnification of (a).

solder contamination accumulated at the head-glass disk interface. When the soft solder accumulated at that interface, it became smeared because it underwent plastic deformation quite easily. The thickness of the smear was less than the head flying height of 0.25 μm (10 μin.).

Recommendation. Although the solder seemed to be adhering quite strongly to the slider, further head-disk interaction could loosen it, thereby damaging the disk. All potential sources of solder should be investigated for possible loose particle generation.

CASE STUDY 7.4: Hard Dust Particle Scratches on Alumina

Problem. A thin-film head exhibited scratch damage of the sputter-deposited alumina layer after testing.

Analysis. Figure 7.14 shows that the scratch was on the sputtered alumina layer. It should be noted that, because the head flies at a shallow angle, called "pitch," with its trailing edge closer to the disk than its leading edge, the edge of the alumina is the closest to the disk. As a result, the scratch is the deepest at the edge of the alumina.

Solution. Because the alumina was scratched, it was evident that the contaminant had higher mechanical hardness. EDXS analysis of the fine debris near the scratches revealed the presence of calcium and silicon, indicating that a calcium silicate dust particle caused the scratch. If the dust particle originated at the head-magnetic disk interface, the particle would have preferentially scratched the disk because the disk is softer than alumina. Because alumina is softer than the glass disk used for fly testing, it was concluded that the dust particle that caused the damage originated at the head-glass disk interface.

Recommendation. Potential sources of dust contamination during fly testing should be investigated.

CASE STUDY 7.5: Handling Damage Prior to or During Assembly of Drive

Problem. A disk drive experienced read/write problems. When it was opened, the thin-film head was found to have chipped alumina. Was the chipping related to the problem?

Analysis. Figure 7.15 shows the location where the sputtered alumina layer was chipped. Media debris, consisting of loose particles from the magnetic medium of the disk, collected over the edge of the chip.

Solution. The fact that flying debris accumulated on the chipped edge indicates the chip was present before the head began flying over the disk. Also, the fact that the damage was on the outside edge of the head indicated that it was due to handling prior to or during assembly of the drive.

Recommendation. While chipping may not be the problem, more care was needed during shipping and handling of the head prior to and during drive assembly. Heads with chipped alumina could scratch the disk.

CASE STUDY 7.6: Use of Electroless Nickel-Coated Steel Parts in Drives

Problem. Too many errors were experienced by a drive during testing, and it was disassembled to determine their cause. The head was found to have accumulated contamination.

Analysis. Figure 7.16 shows the accumulated contamination on the trailing edge. EDXS examination revealed the presence of nickel and iron and, in smaller amounts, calcium and chlorine. Fine scratches were visible on the disk.

Fig. 7.15 Accumulated debris at the fractured edge shows that alumina chipped prior to drive assembly.

Solution. Any contamination that is present tends to accumulate on the head because of the air flow pattern. The presence of nickel and iron indicated that an electroless nickel-coated steel part was shedding particles. The presence of calcium and chlorine could indicate hard water deposits, talcum powder, or glove powder.

Recommendation. All drive components and assembly tools should be checked for the presence of any electroless nickel-coated steel parts, and their subsequent quality should be verified. Strict adherence to clean-room practices is necessary to eliminate calcium- and chlorine-containing contamination.

CASE STUDY 7.7: Handling Damage from Plastic Boxes

Problem. A sampling of heads was routinely studied for contamination buildup after testing. A few heads exhibited contamination on the trailing edge. What was the source of the problem?

Fig. 7.16 Accumulated nickel and iron debris shows an electroless nickel-coated steel part generated particles.

Analysis. Figure 7.17 shows an example of contamination that collected on the trailing edge. The EDXS study indicated that it was probably a hydrocarbon.

Solution. The fact that the contamination was present on the air bearing surface indicated that it originated after testing. Otherwise, the spinning disk would have carried off any contamination present or smeared it. The fact that it accumulated on an edge indicated that it originated during handling. Because it was found to be a hydrocarbon, the plastic boxes used in the shipping and transport of the heads from the disassembly site to the analysis site were regarded as the likely source of contamination. It should be remembered that alumina is much harder than plastic and is capable of digging into it and scraping off particles of plastic during contact.

Recommendation. The tested heads did not experience any contamination problems. Extra care should be taken when inserting and removing heads from the plastic shipping trays.

Fig. 7.17 Contamination on the air bearing surface shows the head did not fly over the disk after it occurred, which suggests handling damage after drive disassembly.

CASE STUDY 7.8: Chipping as the Source of Failure

Problem. A failed disk drive was disassembled, and the thin-film head was found to exhibit signs of chipping. Could the chipping have been the cause of the failure?

Analysis. Figure 7.18 illustrates fracture of the sputtered alumina layer at the trailing edge, where some media debris accumulation is visible at the edges.

Solution. The fracture morphology and stress lines indicated that it was an impact fracture. The head hit a blunt part like a tool or a drive component during assembly. The debris accumulation indicated that the head flew over the disk after the fracture. It should be

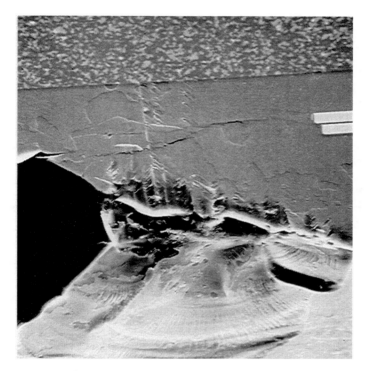

Fig. 7.18 Impact fracture of alumina. Accumulated fly debris showed it occurred before head-disk assembly. Alumina chips could damage the head-disk interface.

remembered that, because the fractured edge is sharp, it tends to dig into the disk media and collect more debris.

Recommendation. Such an extensive fracture can generate alumina chips and can also scratch the disk because of sharp and protruding edges. Extra care is needed during handling and assembly of the heads to prevent this type of damage.

CASE STUDY 7.9: Solvent Stains During Testing

Problem. Following testing, a head exhibited air bearing surface contamination.

Analysis. Figure 7.19 shows the air bearing surface of the head. An EDXS study revealed that the spots were probably hydrocarbons.

Solution. The presence of round spots indicated that they originated after testing. Otherwise, the rotating disk would have smeared

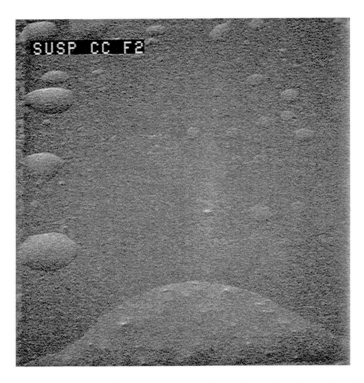

Fig. 7.19 Solvent spots from post-assembly solvent cleaning. Head-disk interaction would have removed or smeared spots present during testing.

the contamination during takeoff and landing. Their round shape indicated that they were solvent spots, or residue left after a cleaning solvent evaporated. Their presence indicated that the head was cleaned with an impure solvent after testing.

Recommendation. Because an attempt was made to clean the heads after testing, no determination could be made concerning contamination buildup during testing.

CASE STUDY 7.10: Sintering Problems in Nickel-Zinc Ferrite

Problem. Nickel-zinc ferrite pads exhibited discoloration on the air bearing surface.

Analysis. Figure 7.20 shows the ferrite surface. The EDXS examination indicated that the gray area had an abnormally high nickel content, whereas the porous area had an abnormally high zinc

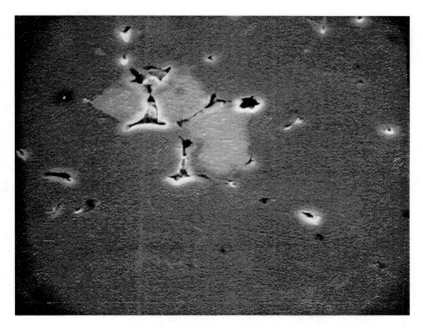

Fig. 7.20 Defective nickel-zinc ferrite. Gray area contains excess nickel; porous area contains excess zinc. Probably caused by a problem with the ferrite sintering process. SEM, 2000×.

content, and the surrounding area contained normal amounts of nickel and zinc.

Solution. The segregation of nickel and zinc indicated that the raw material used in ferrite manufacturing was not blended adequately. As a result, during sintering, a homogeneous composition could not be obtained. These two areas lapped at different rates during lapping and polishing, leading to the appearance of discoloration.

Recommendation. Blending and sintering of the nickel-zinc ferrite material should be optimized to obtain uniform composition.

CASE STUDY 7.11: Particle Generation from a Static Ground Spring

Problem. A crashed drive was disassembled to determine the cause of failure.

Fig. 7.21 Accumulated silver and graphite on the head, caused by a static ground spring particle. SEM, 1000×.

Analysis. Figure 7.21 shows the contamination buildup on the head, which was found to be iron and silver. The accumulated debris was removed and studied with X-ray diffraction, which showed peaks corresponding to graphite, silver, and gamma iron oxide.

Solution. Because EDXS (with a beryllium window) does not reveal the presence of carbon, the presence of graphite could only be identified from X-ray diffraction studies. The iron peak was caused by the iron oxide media debris generated by the disk during the head-disk crash. An investigation of the drive components showed that the static ground spring was made of 75% silver and 25% graphite. Particles generated by the spring was the probable cause of the crash.

Recommendation. It was recommended that the quality of the static ground spring be improved.

Appendix 7.1

So That's How They Keep The Bits On The Disk!*

The information-storage business is going through a rapid development period in its history, with ever-increasing storage densities accompanied by a strong emphasis on improving product quality and reliability while reducing costs. As Professor Genichi Taguchi points out, "Quality is the loss imparted to society from the time a product is shipped."

Making a product well the first time is one of the best and most effective ways of reducing costs. Producing a better quality disk drive, possibly at a lower cost, is the only way of staying ahead of the severe competition in the computer peripherals business. A better quality drive is obtained by improving the quality of drive components such as the media and the read/write head.

The head/disk assembly (HDA) is technologically the most challenging part of a disk drive. During drive operation, if the read/write head (usually encased in a hard ceramic) inadvertently touches the soft media on the disk, it will scrape off the media, and the stored information will be lost forever. Such a catastrophic loss is called head crash. Materials testing procedures that have been done to produce and certify a high-quality rigid disk with optimum parametrics and longevity that will prevent (or at least delay) head crash are described below.

---

* A.S. Brar and P.B. Narayan, *Research & Development*, Oct 1987, p 100-104. Reprinted with permission.

Disk Manufacturing

A rigid disk (i.e., a disk with an aluminum alloy substrate) is
mechanically more stable than a flexible disk and can accommodate
a higher number of tracks per inch. For this reason, rigid disks are
used in high-bit-density applications.

The current trend in the storage peripherals industry is to increase the bit density while decreasing the physical size of the disks.
Reducing the disk size makes it possible to design a drive that uses
a smaller motor, consumes less power, generates less heat, and has
other associated advantages when compared to conventional drive
designs.

Increasing bit density requires magnetic coatings of 0.5 µm or less
and read/write head flying heights above the disk of 0.25 µm or less.
With ever-decreasing media parameters, it becomes more important
to maintain tight control on the physical properties of the disk
materials and to have a thorough understanding of those properties.

Although plated and sputtered metal films look promising as
magnetic media materials, most media available today use gamma
iron oxide or a cobalt-modified version of it as the data storage
material. Acicular gamma iron oxide particles (about 0.6 µm long) are
milled, along with surfactants and a binder resin, and mixed with
milled alumina (for wear resistance) to form a magnetic paint.

At each stage of the disk coating process extensive testing is done
to maintain oxide quality. The schematic in Fig. 1 shows typical tests
and at what stages they are performed.

The polished and diamond-turned aluminum alloy substrate is
then coated with the magnetic paint using a spin-coater. The iron
oxide particles are oriented circumferentially along the disk using a
suitable magnetic field. Finally, the disk is baked to cure the binder
resin and then polished to obtain the required surface finish and
magnetic coating thickness (Fig. 2).

In early disk drive designs, heads were placed above the disks only
after the latter reached their maximum speed and were taken out of
the module before the drive was turned off. In more-recent Winchester drive designs, the head rises from the disk, flies above it, supported by the aerodynamic air bearing produced by its rotation, and
then lands on the disk when the drive is turned off.

This design requires that a lubricating fluid be applied to the rigid
disk surface before it is placed in the module. The lubricant reduces
the wear rate in the head landing/takeoff zone.

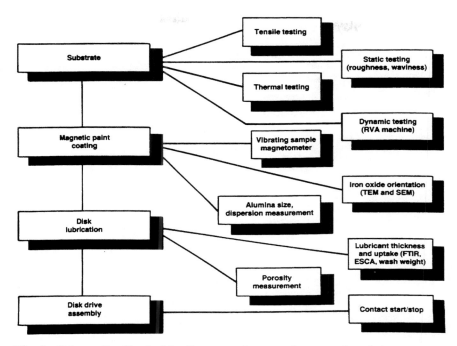

Fig. 1 Schematic of typical testing procedures and stages of implementation.

Disk Substrate

A good disk requires a high-quality substrate. A finished disk undergoes a lot of dynamic and static mechanical testing to ensure that it is compatible in a head/disk assembly with a flying height of about 0.25 μm. The bulk properties of the disk measured in the static mode are the elastic and thermal expansion coefficients.

Because the disk is rotating at 3600 rpm, it is important to know the modulus of elasticity of the substrate. Elastic constants such as Young's modulus, modulus of rigidity, and Poisson ratio are determined by a tensile-testing machine.

Because the thermal expansion coefficient for the 5086 aluminum alloy is about 14 μm/m · °F, a change of a few degrees Fahrenheit in the ambient temperature will lead to the relative change of location of a track compared to its neighbor. During the drive design process, the linear thermal expansion of the disk material should be matched

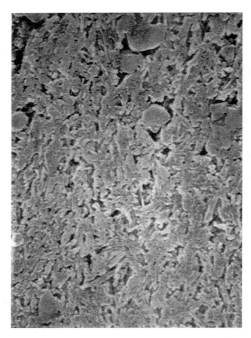

Fig. 2 SEM micrograph of oxide paint shows the surface alumina (large particles) that is used to reduce wear at the disk drive head/disk interface.

with that of the head/arm material. The thermal expansion coefficient of the substrate is determined by a thermomechanical analyzer.

Because of the critically small media parameters, the substrate surface finish is extremely important. High-purity 5086 aluminum alloy is widely used as substrate material. The blank substrate is chamfered, thermoflattened, and diamond turned. In some cases, the substrate is then polished and textured.

Static testing of the surface finish involves surface profilometer traces (the average roughness should be less than 0.01 μm) and measuring the radius of curvature and radial waviness. Dynamic testing is much more complex and involved.

Typical dynamic properties such as circumferential displacement and circumferential acceleration are measured by an RVA (run out, velocity, and acceleration) machine. RVA machines encompass much innovation and complex technology and, hence, are not readily available in the market.

Dynamic properties of the disk substrate must continue to improve to meet the stringent requirements of future high-density disk drives. This can only be done after significant improvements are made in dynamic properties measurement technology.

It is clear that a lot of labor is involved in preparing a substrate (e.g., diamond turning, machining, and polishing). Also, aluminum alloys tend to contain second-phase (AlFeSiMn) particles that create missing bits and data errors during the read/write process. Therefore, very high-purity aluminum alloys are preferred for high-density disks. An additional problem with aluminum is its tendency to corrode.

Because of these cost and corrosion considerations, much work is being done to find an alternative to aluminum alloys. Glass, ceramics, and plastics are seriously being considered. It may be assumed that the next significant advancement in disk manufacturing technology will be in the substrate material.

Magnetics of the Media

Just as the study of substrate static and dynamic properties is essential for ensuring mechanical functioning of the head/disk assembly without danger of head crashes, the study of media magnetics is essential to ensure electrical performance of the head/disk assembly. Magnetic properties are measured by a vibrating sample magnetometer (VSM) that vibrates a 0.5-cm-diam specimen of the disk at about 80 cps in a magnetic field of 10,000 oersteds.

The oscillating magnetic field of the specimen induces an alternating electromotive force (emf) in the detection coils. This emf is compared with the emf generated by a reference specimen. The difference between these two emf values is proportional to the magnetic moment of the specimen.

Many VSMs are available with computerized controllers that also record the coercivity, saturation magnetization, remanent magnetization, and squareness values. The disk magnetic properties are measured in both circumferential and radial directions and the orientation ratio (or the ratio of remanent magnetizations in the two directions) then can be calculated (Fig. 3).

The amplitude, resolution, and signal-to-noise ratio of a recorded signal during the read/write process improve with the orientation of

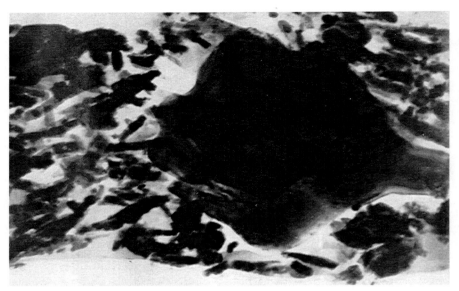

Fig. 3 Circumferential TEM image of an oxide layer shows iron oxide particles oriented circumferentially. Large particle is a piece of alumina.

iron oxide on the disk. The orientation can be studied by TEM or SEM techniques.

The TEM technique involves cross sectioning the disk and studying the orientation throughout the thickness of the magnetic coating. To study the iron oxide orientation, it is important to obtain TEM micrographs along both circumferential and radial directions on the disk, because the iron oxide particles are oriented along the circumferential direction. This technique is difficult to use and is time consuming. Thus, it generally is used during the coating development process to optimize the magnetic field required to obtain the desired orientation.

The SEM technique involves plasma ashing some of the binder resin on the disk and obtaining SEM micrographs at specified magnifications (such as 3000× and 6000×). SEM micrographs reveal the orientation of the oxide particles that are present only at the top surface of the coating. However, because it is a relatively simple technique, an SEM is the most commonly used instrument for monitoring iron oxide orientation.

Load-bearing ceramic (mainly alumina) particles are included in the magnetic coating for wear resistance. The role of alumina is not completely understood, but it generally is believed that it cleans debris particles off the read/write head, thus preventing or delaying head crashes. Obviously, only alumina that is present at the head/disk interface will be effective.

One of the disadvantages of alumina is that when it is agglomerated it may contribute to missing bits, extra bits, and head crashes. Thus, it is important to monitor its distribution on the top surface of the disk.

Alumina distribution is monitored by plasma ashing the disk to burn away a part of the resin binder and then obtaining SEM images of the remaining surface. From these images, the alumina size distribution is calculated and a typical histogram generated.

Surface Porosity and Lubrication

In Winchester disk drives, as previously stated, the head flies above the disk during the read/write process on an aerodynamic air bearing, but takes off from and lands on the disk at the beginning and end of this operation. To reduce wear from head takeoff and landing a fluorocarbon-type lubricating fluid is applied to the disk either by dip-coating or by a spray and buff technique.

If the lubricant coating is too thin, the head will wear away the media. If the coating is too thick, stiction will occur, causing the head to adhere to the lubricant. Both of these conditions can cause a head crash.

Ideally, the lubricant should be about 0.005 µm thick, and the media should contain numerous pores in which the lubricant can be stored (Fig. 4). During disk drive operation, some of the lubricant may be spun off of the disk surface because the disk rotates at 3600 rpm. When that happens, the lubricant should be replenished from pores that are acting as lubricant reservoirs.

During the development of a coating formulation, porosity is investigated by a TEM study of media cross sections. However, an SEM is used to routinely monitor porosity. Porosity measurement is primarily a qualitative technique, and the SEM micrographs are retained and cataloged to check for any gross changes in the number and shape distribution of pores.

Lubricant uptake by the disk may be inferred from the wash/weight method in which the lubricant is washed off of the disk

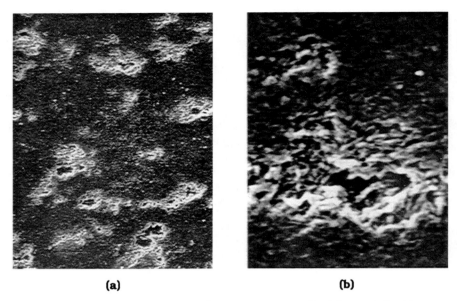

(a)   (b)

Fig. 4 Pores in the oxide material act as reservoirs for lubricant that is
used at the head/disk interface. Micrograph, obtained using a low-voltage
SEM, shows several pores (a) and a close-up of a pore area (b).

and weighed. This weight gives the total weight of the lubricant, both
on the surface and in the pores.

Fourier transform infrared spectroscopy (FTIR) also is used to
monitor the coating lubricant supply, and electron spectroscopy for
chemical analysis (ESCA) can be used to measure the actual surface
lubricant thickness, but this surface technique does not indicate the
amount of lubricant trapped in the pores. Thus, a combination of
wash weight, FTIR, and ESCA provides information about the actual
lubricant thickness and the amount of lubricant present in the
pores, thereby enabling one to predict how long the lubricant will last
on the disk.

Thus far, disk reliability and performance in terms of testing and
verification at each step of the disk manufacturing process have been
described. As a final test, a representative sample from each lot of
disks is tested for accelerated wear in the unlubricated and lubricated states, using a contact start/stop tester. This test indicates

how various head/disk assembly parts work as a unit. Any erratic behavior of the disks in the tester warrants a careful repetition of the various testing procedures described above.

As the bit density and performance standards of disk drives improve, system parameters such as the media thickness and the head flying height must decrease. This leads to a need for a more thorough and critical testing of parts during various manufacturing states to ensure that drive components function flawlessly.

Chapter 8

Plating and Sputtering

Thin films play a crucial role in the microelectronics industry because they can be manufactured while maintaining precise control of materials properties. The two most popular thin-film manufacturing methods are plating (electroplating and electroless plating) and sputtering. See Ref 1 for more information on thin-film deposition technologies.

Plating is a well-studied process that produces thin films with good control of composition, thickness, and microstructure. It is used to manufacture most high-performance read/write heads (called thin-film heads) and some of the thin-film disk media (called plated media). Case studies dealing with this type of precision and high-quality plating are presented below. Plating is also used to provide corrosion resistance and to reduce contamination. These applications and corresponding case studies are presented in Chapter 3 on protective coatings.

Sputtering is the preferred vacuum or nonaqueous technique to deposit thin films, compared with chemical vapor deposition and thermal evaporation, because it provides superior control over microstructural features such as composition and thickness, which determine the resulting materials properties.

Plating is performed in aqueous solutions, which causes corrosion resistance of the plated films to be a concern because of the problem of removing trace amounts of water. Sputtering is a dry technique and as such does not pose water retention problems. Sputtered films are more porous than plated films because of argon entrapment. Of the two, plating is the relatively inexpensive process.

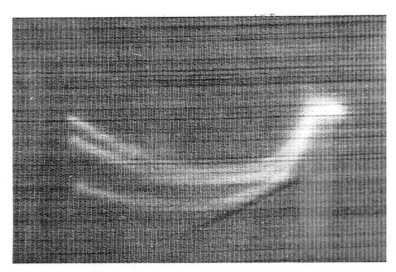

Fig. 8.1 Organic fiber placed on a substrate during carbon deposition, which produced a shadowing effect that gave the impression of flaking carbon. 2900×.

Appendix 8.1 discusses the effects of common contaminants on the plated surface finish. In this case, Permalloy (NiFe) electroplated film was used in the thin-film heads.

CASE STUDY 8.1: Fiber Contamination Leading to Flaking

Problem. In the manufacture of plated media, the aluminum alloy disk substrate is plated with an electroless nickel film approximately 5 μm (200 μin.) thick, and then plated with a 40-nm (1.58 μin.) thick cobalt-based alloy, e.g., CoNiP or CoP. A 30-nm (1.18 μin.) thick carbon overlayer is sputter-deposited to provide corrosion and wear resistance. During optical inspection, the carbon-coated and lubricated disk exhibited discoloration that appeared to be flaking of the carbon film (see Fig. 8.1).

Analysis. Auger electron spectroscopy (AES) indicated the presence of carbon in the discolored region, but the carbon was thinner in the discolored region than in the surrounding region.

Solution. The Auger results, coupled with a closer study of the region, indicated that a cloth or paper fiber was present beneath the carbon layer. As the fiber moved during the deposition process, it

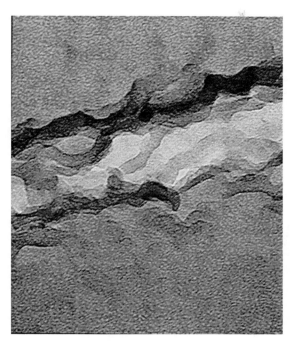

Fig. 8.2 Carbon overcoat showing inadequate coverage of carbon at the texture lines, which makes the carbon weak and prone to fracture. TEM, 30,000×.

gave the impression that the overcoat was flaking. After plating, the disk was dried and carried into the sputter chamber, and this drying and handling operation introduced the use of a cloth that generated fiber contamination.

Recommendation. The contaminating cloth was removed from the operation, and more attention was paid to reduce the handling and waiting time between the plating and sputtering processes.

CASE STUDY 8.2: Nonuniform Coverage of Carbon on Rough Surfaces

Problem. The hard disk substrate is provided with concentric grooves (called texturing) to optimize magnetic and tribological properties of the disk. During texture optimization experiments, a disk with a very rough texture experienced unusual carbon buildup on

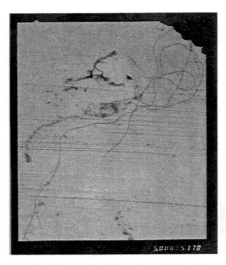

Fig. 8.3 Fiber contamination adhering strongly to the carbon overcoat, causing occasional contact with the head. TEM, 16,000×.

the head and an abnormal increase in friction when the disk was wear tested with a normal head.

Analysis. Auger electron spectroscopy confirmed the presence of carbon on the head slider surface. The magnetic layer on the disk was removed by acid dissolution, and the carbon layer was studied with transmission electron microscopy (TEM).

Solution. Figure 8.2 is a TEM micrograph showing the inadequate coverage of carbon at the texture lines, which renders the carbon weak and prone to fracture and removal during wear testing. The carbon buildup on the slider increased friction at the head-disk interface.

Recommendation. Texturing generally improves magnetic and tribological properties of the disk. However, it cannot be assumed that any texture is better than no texture. Texturing needs to be optimized by careful study (Ref 2).

CASE STUDY 8.3: Fiber Contamination Leading to Flight Interference

Problem. During head-disk interference testing, one lot of disks exhibited an abnormally high number of "hits," indicating that

Fig. 8.4 Blister formed under a cobalt alloy plating due to fluid entrapment in substrate pores. SEM, 100×.

during the flight, the head met with many disk imperfections such as protrusions or particulate contamination.

Analysis. The scanning electron microscope (SEM) examination of the disk did not show particulate contamination. The magnetic layer was dissolved by acid dissolution, and the carbon overcoat layer was studied with TEM.

Solution. Figure 8.3 is a TEM micrograph of the carbon, indicating organic fiber contamination. The selected area electron diffraction pattern (SAEDP) revealed an amorphous structure, confirming that the fiber was organic in nature. The micrograph also shows that the fiber did not just fall on the disk, but adhered to it strongly. As a result, the fiber did not "fly away" from the rotating disk and led to the high number of "hits" encountered in this test.

Recommendation. The post-carbon deposition processing was carefully studied to ensure that there was no contact with any fiber generator such as a cloth or a cotton cleaning swab.

Fig. 8.5 Partially open blister. SEM, 100×.

CASE STUDY 8.4: Surface Porosity, Fluid Entrapment, and Blisters

Problem. A die-forged aluminum alloy was coated with a cobalt-phosphorus alloy by electroless plating. Blisters subsequently developed on the surface.

Analysis. Figure 8.4 shows a blister growing on the plated surface. A partially open and fully open blister are shown in Fig. 8.5 and 8.6, respectively. Evidence of corrosion salts is visible in Fig. 8.6. When all corrosion products were removed, the substrate was clearly visible (Fig. 8.7).

Solution. Surface pores entrapped the plating solution, which could not be washed off. The moisture slowly reacted with the

Fig. 8.6 Blister that is fully open, showing corrosion products. SEM, 1824×.

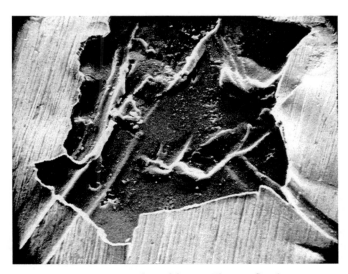

Fig. 8.7 Substrate porosity under a blister. Plating fluids were entrapped in the pores, causing blister formation. SEM, 380×.

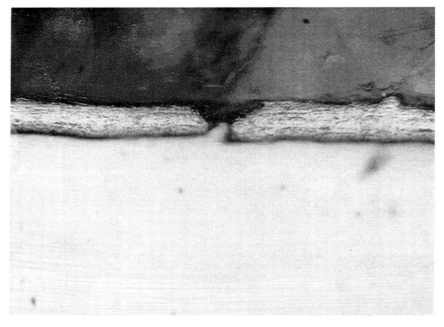

Fig. 8.8 Cross section of a part showing cracks in thicker platings. Thicker platings tend to charge more hydrogen into the steel substrate, potentially causing hydrogen embrittlement. 500×.

aluminum substrate and the cobalt alloy, forming a corrosion product that encouraged growth of the blister.

Recommendation. The substrate should be free of porosity. Excessive humidity should be avoided during forging of the substrate, to minimize the formation of internal voids. Additionally, the material could undergo solid-state degassing before forging.

CASE STUDY 8.5: Fracture Due to a Rough Surface and Thick Coating

Problem. One lot of leaf springs, made of plain-carbon steel coated with zinc or cadmium, exhibited premature failure.

Analysis. Figure 8.8 is a cross section of a failed spring showing that the plating broke. Figure 8.9 shows the rough surface of the substrate.

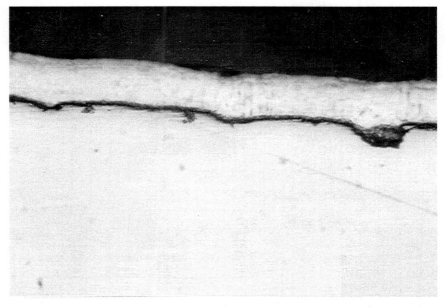

Fig. 8.9 Cross section of a failed spring exhibiting extreme surface roughness. Surface roughness, machining marks, and other surface defects can act as stress concentrators. 500×.

Solution. The rough substrate surface could produce stress concentrators, which could lead to fracture of the spring. The failure became worse as the coating thickness increased. In general, as the coating thickness increases, the amount of hydrogen charged into the substrate also increases.

Recommendation. The substrate surface must be smooth and free of machining marks and other stress concentrators. Thinner coatings are preferred. The part should be annealed at 200 °C (390 °F) within 1 h of plating to remove hydrogen, because hydrogen causes steel to become brittle, by hydrogen embrittlement.

References

1. R.F. Bunshah, *et al.*, *Deposition Technologies for Films and Coatings*, Noyes Publications, 1982

2. E.M. Simpson, P.B. Narayan, G.T.K. Swami, and J.L. Chao, "Effect of Circumferential Texture on the Properties of Thin-Film Rigid Longitudinal Media," *IEEE Trans. Magn.*, Vol 23, 1987, p 3405

Appendix 8.1

Permalloy Plating Imperfections—
Effect of Contamination*

Imperfections in the permalloy film, used in read/write heads, interfere with domain switching, leading to noise problems such as glitch. In the present study, several types of permalloy surface defects were evaluated at various thicknesses to understand their origin and the morphological changes with plating thickness. The effect of the presence of contamination (both particulate and hydrocarbon) and other substrate surface defects on the growth and morphology of permalloy were investigated using scanning electron microscopy (SEM), transmission electron microscopy (TEM), and energy-dispersive X-ray spectroscopy (EDXS).

Introduction

Plated permalloy with approximately 82% nickel and 18% iron is the soft magnetic material that is widely used in inductive thin-film heads for disk drive applications (Ref 1). During the read-back process, domain wall movement is adversely affected by material defects in permalloy such as inclusions, compositional differences, differential stress states, and other nonuniformities that could act as domain pinning centers (Ref 2). Such impediments to domain wall movement cause noise or glitch in the read-back signal.

To manufacture heads with minimum noise, permalloy should be as homogeneous as possible. The homogeneity is affected by nuclea-

---

* P.B. Narayan and S.C. Herrera, StorageTek, Louisville, Colorado

tion and growth processes of the film, which are, in turn, affected by the substrate surface preparation and the plating process itself.

The present work deals with the contamination-induced defects found in permalloy and their morphology. This study is expected to provide help in understanding the origins of defects by observing the morphology of the surface of the plated film, with the ultimate objective of eliminating the material defects.

Experimental

Because the basic objective of the present work is to understand how films grow on various contaminants, the substrate was exposed to the common contaminants before depositing permalloy films.

To fabricate the bottom pole of the read/write head, alumina is sputter deposited on a composite ceramic substrate such as alumina-titanium carbide. The substrate is then lapped and polished to obtain the required surface smoothness. A permalloy seed layer (nominally 0.1 μm thick) is sputter deposited to provide the electrical conductivity required for electroplating. To fabricate the top pole, the seed layer is deposited on a cured photoresist layer.

Several kinds of permalloy surface defects were studied to determine their exact origin by chemically etching away permalloy at the areas of interest in a controlled manner and investigating with optical microscopy, electron microscopy and energy dispersive X-ray spectroscopy.

Results and Discussion

Based on the present study, four types of common contaminants and substrate irregularities were identified for discussion.

Particulate Organic Contamination

Particulate organics are perhaps the most frequently observed contaminants in manufacturing areas and clean-room environments. Cloth fibers, paper debris, air filter materials, plastics, and common adhesives are some examples of particulate organics.

Sputtered permalloy, used as a seed layer for permalloy plating, is surface activated with a chemical etchant to remove the oxide layer

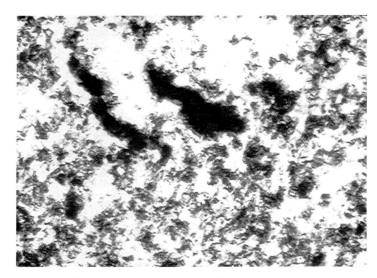

Fig. 1 Sputtered permalloy seed layer with particulate organic at the top left side. TEM, 100,000×.

so as to improve the adhesion between the seed layer and the plated film. Figure 1 is a TEM micrograph of the seed layer, with a fibrous contaminant visible toward the upper left side. The film around it appears to be porous and defective. Figure 2 is a SEM micrograph of a pitting defect observed on a surface-activated seed layer. Particulate organic contamination (EDXS reveals it to be a hydrocarbon) is evident at the center of the defective area.

The film formed around the contamination was thin, porous, and highly stressed. As a result, that area was chemically very reactive and readily reacted with the etchant. The plated film formed on such a defective area will not be homogeneous.

Figures 3 and 4 are optical and SEM micrographs of a plated film surface showing a defect due to a particulate organic contaminant. The contamination at the center caused nonuniform growth of permalloy. Immediately surrounding the defect, the film was substantially thin, and the area a few micrometers away showed inhomogeneous growth of permalloy.

Figure 5 shows the defective area after half of the 2-μm thick permalloy was chemically etched away, and Fig. 6 after 90% of the

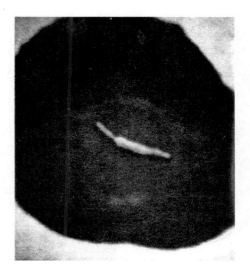

Fig. 2 Pit on seed layer after surface activation. Notice particulate organic at the center.

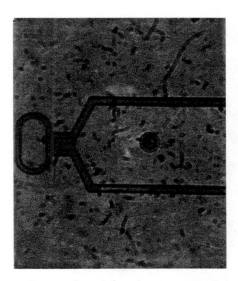

Fig. 3 Plated permalloy surface defect due to particulate organic contaminant. Optical, 1300×.

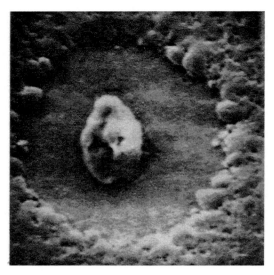

Fig. 4   SEM micrograph of the defect in Fig. 3.

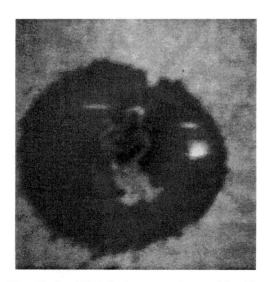

Fig. 5   Defect (Fig. 3) after 50% thickness etching of the film.

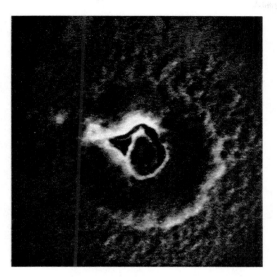

Fig. 6 Defect (Fig. 3) after 90% thickness etching of the film.

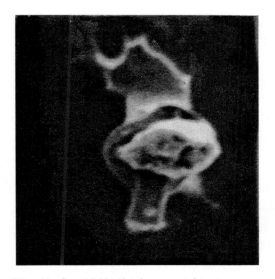

Fig. 7 Defect (Fig. 3) after 100% thickness etching.

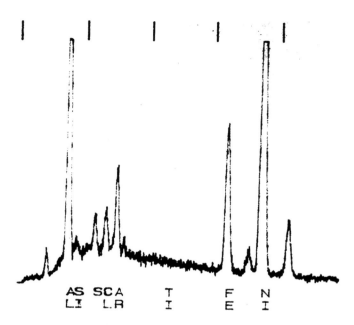

Fig. 8 EDXS spectrum of the contaminant (Fig. 3) showing chlorine and sulfur in the organic contamination.

film thickness was etched away. Figure 7 is the same area after 100% etching.

Because the contamination adhered to the surface even after the complete removal of the seed layer, it must have been present on the substrate beneath the seed layer. Figure 8 is the EDXS compositional spectrum showing the presence of chlorine and sulfur in the foreign particle, indicating particulate organic contamination probably related to human handling.

The organic contamination is electrically nonconducting. It also facilitates the formation of a hydrogen gas bubble. Figures 1 and 2 indicate that the seed layer formed around the contaminant is defective. As a result, the plated film is much thinner and rougher in the surrounding area.

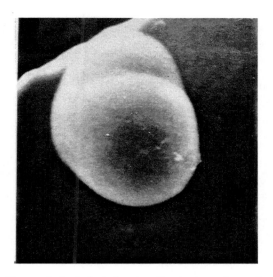

Fig. 9 Permalloy nodule defect due to silicate-type dust. Optical, 1300×.

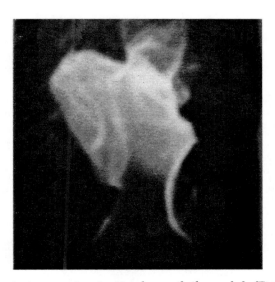

Fig. 10 Silicate-type contamination beneath the nodule (Fig. 9).

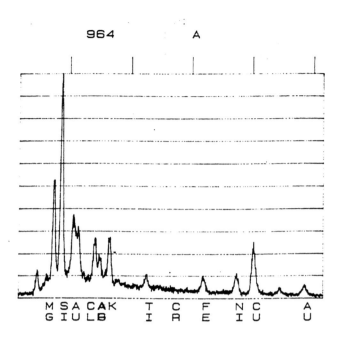

Fig. 11 EDXS spectrum of the particle (Fig. 10) showing silicon, titanium, and magnesium.

Silicate-Type Particulates

Calcium, magnesium, aluminum, and titanium silicates are normal constituents of atmospheric dust, cement, sheetrock, and other construction materials. Some of these elements are also present in deodorants, talcum powder, and other cosmetic materials.

Figure 9 is an example of plating nodules observed on the surface of plated permalloy, and Fig. 10 shows the silicate-type contaminant found beneath the nodule. Figure 11 is the EDXS spectrum showing the presence of silicon, magnesium, and titanium.

Silicate-type particles are much better electrical conductors than organics. Because the particle is sticking out of the substrate, the current density is abnormally high near the particle, and the plating produces a nodule very rapidly.

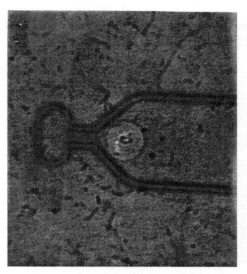

Fig. 12   Surface defect due to film-like hydrocarbon. Optical, 1300×.

Film-Like Hydrocarbon Contamination

Film-like hydrocarbon contamination is perhaps the most insidious of all contaminants, because it is present in all environments and it is very difficult to detect. Machining oils, vacuum pump oils, and organic compounds that evolve due to outgassing are some of the sources of film-like contamination.

Figure 12 shows a permalloy surface defect, and Fig. 13 shows the same defect after partial etching. Figure 14 shows the surface defect after complete etching. The beading structure of the film indicates the probable presence of hydrocarbon film-like contamination on the substrate. The contamination layer was too thin to be detected by EDXS. The film is thinner and rougher over the contaminated areas because of an inhomogeneous seed layer and nonuniform electrical conductivity during plating.

Surface Irregularities

Improper surface preparation (e.g., lapping and polishing), insufficient cleaning, and unwanted chemical reactants on the surface are some of the causes of pitting and other roughness irregularities.

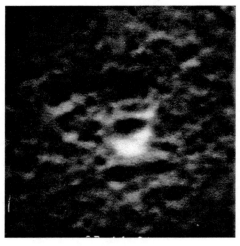

Fig. 13   Defective area (Fig. 12) after 90% thickness etching of the film.

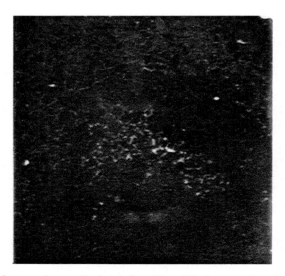

Fig. 14   Substrate beneath the defect (Fig. 12) showing beading of the seed layer. Probable presence of hydrocarbon film.

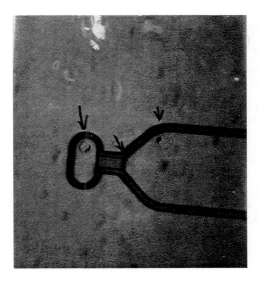

Fig. 15 Surface defect due to pitted substrate. Optical, 1300×.

Figure 15 is an optical micrograph of a defective area, indicating a higher roughness than the surrounding permalloy surface. Figure 16 shows the same area after complete permalloy etching, showing pitting of the substrate alumina.

A pitted surface produces a rougher seed layer and consequently a rougher plated film because of the poor throwing power of permalloy plating baths.

Summary

Various defects observed on plated permalloy were investigated by controlled chemical etching and SEM/EDXS. Particulate organics, silicate-type dust, film-like hydrocarbon contamination, and surface irregularities create characteristic surface defects in permalloy. For desirable and reliable recording properties, it is essential to understand the origin of defects so as to eliminate them. An ultra-clean manufacturing environment is essential for producing high-quality heads.

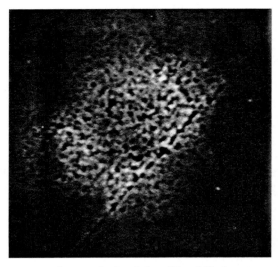

Fig. 16 Substrate underneath the defect (Fig. 15) showing pitted alumina substrate.

Acknowledgments

The authors thank G. Baubock and R. Muto for their support and encouragement.

References

1. L.T. Romankiw, "Thin Film Inductive Heads: From One to Thirty One Turns," *Proc. Symposium on Magnetic Materials, Processes and Devices,* L.T. Romankiw and D.A. Herman, Jr., Ed., The Electrochemical Society, 1990, p 39-53 and other papers therein.
2. K.B. Klaassen and J.C.L. van Peppen, "Barkhausen Noise in Thin-Film Recording Heads," *IEEE Trans. Mag-26,* Vol 1697, 1990

Chapter 9

Ball Bearings

In a hard disk drive, besides the head-disk interface, the most challenging component from the point of view of quality and reliability is the bearing. Because ball bearings are the bearings of choice for use in disk drives, this chapter will focus on their use and factors contributing to their possible failure. As the access speeds and data transfer rates continue to increase, the complexities of ball bearing manufacturing and their failure modes are increasing.

A bearing is considered to have failed when either of two criteria have been met: (1) the bearing specification has not been met (e.g., vibration, noise, and torque level) and (2) microcontamination has been generated in the work environment.

From an applications point of view, it is advantageous to discuss ball bearings as spindle bearings, linear actuator bearings (including rails), and rotary actuator bearings.

9.1 Spindle Bearings

Hard disks are stacked on a spindle that rotates with the aid of a ball bearing at high rotational velocities during read/write operations. In first-generation drives, the rotational velocity was 2400 revolutions per minute (rpm). In second-generation drives, including 3380 drives, velocity increased to 3600 rpm. In the current generation, it is 5400 and 7200 rpm. Higher velocity provides not only higher data transfer rates and shorter access times, but also larger read signals, because in an inductive read head, the read voltage is proportional to the rate of change in the magnetic flux.

In modern drives, the demand for higher areal storage densities restricts the head flying height to as low as 100 nm (3.94 µin.) above the rotating hard disk. This low flying height requires that the tolerable noise and vibration levels of the bearings be further decreased.

9.1.1 *Materials*

For a general discussion of bearings and their failure modes, the reader is referred to *Metals Handbook* (Ref 1). However, a brief discussion of the relevant and salient points of bearing design and materials selection is presented below.

In a ball bearing, the rolling elements are positioned between the raceways and are maintained in their positions by a cage. The relative motion is permitted by the rotation of the balls. A suitable lubricant in the bearing prevents metal-to-metal contact between the balls and races. The bearings are sealed to prevent atmospheric contamination from entering the ball-race interface.

Because bearing balls and races should have high hardness to provide acceptable wear resistance, hardened and tempered martensitic steels are the preferred materials for these elements. Type 52100 steel containing 1.5 wt% Cr and 1.00 wt% C, with a hardness of 60 HRC, has been the standard material for balls and races. In moderately corrosive or humid environments and in applications such as disk drives, in which microcontamination caused by corrosion is highly undesirable, type 440C stainless steel may be a suitable alternative. However, type 440C has substantially lower hardness and about 50% lower fracture toughness than 52100 steel, and thus the dynamic load capacity of type 440C bearings is lower.

In special applications such as aircraft engines, M50 tool steel and the carburizing alloy M-50NiL are used as the ball material for improved performance. While M50 and M-50NiL alloys are more costly than 52100 or type 440C steels, they may be considered for use in future generations of disk drives.

Deposition of hard coatings such as titanium nitride on steel components is being investigated as a means of increasing the life and quality of the bearing. Hybrid ceramic-steel bearings, all-ceramic bearings with liquid lubrication, and solid-lubricated ceramic bearings are among the materials-related innovations aimed at improving bearing performance. Sophisticated analytical techniques such as X-ray CT scanners, eddy current measurements, and ultrasonics are being used to study voids, inclusions, and other defects in the starting materials.

For bearing cages, brass, plastic, and stainless steel are used. In the authors' experience, brass is not an acceptable choice, because it wears easily and is chemically reactive with many of the lubricant constituents.

Technologically, lubrication is perhaps the most complex area of bearing design. Some of the requirements of a good lubricant are: (1) minimum outgassing or evaporation at the bearing working temperature, which is considerably above room temperature, (2) materials stability at working temperatures, (3) no chemical reaction with the bearing materials, (4) easy availability in relatively pure form without particulate or hydrocarbon contamination, (4) non-hygroscopic in nature because of the corrosive action of moisture on bearing materials, (5) appropriate viscosity so as not to allow depletion from certain areas, particularly in special applications such as the rotary actuator.

9.2 Ceramic Bearings

In the machine tool, aircraft, and transportation industries, very high operating speeds and extreme temperatures are encountered, and a new type of bearing, the ceramic or hybrid bearing, is being designed for these applications (Ref 2, 3).

In hybrid bearings, the balls are made of ceramic instead of metal. Sintered and hot isostatically pressed (HIPed) silicon nitride (Si_3N_4) is the optimum ceramic for bearing materials, even though the Hip process causes shrinkage and distortion. Alumina and yttria, common additives to silicon nitride, provide higher strength than magnesia additives. The ceramic components of these bearings are produced in near-net shape because machining from the bulk material involves costly diamond-paste grinding.

The chief advantages of hybrid bearings include longer wear and fatigue life because silicon nitride does not gall against metals, is more tolerant of marginal lubrication, and generates less heat because of its higher elastic modulus. They have a higher speed and acceleration capability because the ceramic used has 50% less density than steel. Also, this lower density reduces centrifugal force on the outer race, thereby lengthening the bearing life. Higher temperature capability and improved corrosion resistance are also possible, because of the chemical inertness of the ceramic.

Hybrid bearings are nonmagnetic, which reduces interference in sensitive instrumentation. They also provide electrical insulation, which prevents electrical arcing damage. These bearings have the ability to operate at a higher thermal gradient for a given internal clearance because of their lower coefficient of thermal expansion, and they offer greater precision because of the higher mechanical stiffness afforded by their higher elastic modulus and lower thermal expansion.

In all-ceramic bearings, both balls and races are made of ceramic material. The use of these bearings is being considered for turbomachinery, furnace mechanisms, etc. Research is under way to eliminate the use of external liquid lubricants with them by introducing solid lubricant into the ball and race material itself. With no liquid lubricant this bearing will be convenient in vacuum equipment and other similar applications.

The future generation of disk drives will be able to capitalize on these advances made in hybrid and ceramic bearings.

9.3 Premature Failure

It is important to remember that a bearing does not have an infinite service life. When the bearing is loaded, elastic deformation occurs at areas of contact between the balls and the races. However, beneath the areas of contact, high shear stresses and low tensile stresses are created, which lead to fatigue cracks and material disintegration (Ref 1). This fatigue failure occurs even under optimum conditions.

It is estimated that only 10% of all bearings function satisfactorily until the end of their normal lifespan (Ref 4). This premature failure of bearings is the focus of this discussion.

The typical signs of bearing failures are (1) audible noise, indicating imminent total failure; (2) loosening of the bearing on the shaft; (3) vibrations in the drive; (4) stiffening of the bearing, requiring high rotating torque; (5) numerous off-track errors; and (6) the generation of microcontamination in the drive.

One of the major causes of premature failures is lubrication problems, e.g., whether it is consistent with the bearing rating or whether conditions of over- or under-lubrication exist. The presence of metallic/hydrocarbon contamination in the lubricant before or during assembly is also a concern.

Failure may also be caused by misalignment of the bearing during preload, or use of improper/substandard cage material, as well as use of too high a preload. Electrical arcing may occur across the ball and race caused by use of a nonconducting lubricant. Additional causes of failure include too much vibration during transport or operation and the presence of contamination caused by ineffective sealing.

9.4 Examination of Failed Bearings

Before a failed bearing is examined, as much information as possible about its history and use should be obtained. Some of the questions that should be asked are:

Why is the bearing considered a failure?

Were any other problems noticed, such as overheating, noise, vibration, or rattling?

Was the bearing exposed to high-temperature or high-humidity environments?

Was there any external shock?

Based on the information these questions elicit, a list of possible causes of failure should be compiled, and failure analysis should be used to systematically eliminate all possible causes until the source of failure is determined.

9.4.1 *Failure Analysis Methodology*

To adequately assess failure, the following procedures should be followed:

1. Examine the exterior of the bearing for any particulate or hydro-carbon contamination. Photograph the exterior of the bearing and any contamination, if present. Open the bearing carefully and inspect the lubricant. Notice the color, consistency, and distribution of the lubricant. Did it accumulate in certain areas? Did it change color (generally to dark brown or black if wear occurred and the wear debris has oxidized)?

2. Mark the location of the inner and outer races with respect to the spindle before the bearing is completely dismantled. Wash off the lubricant with freon or a suitable solvent. Filter the particles and identify them using scanning electron micros-copy/energy-dispersive X-ray spectroscopy (SEM/EDXS). Examine the lubricant with Fourier transform infrared spec-

troscopy (FTIR) and identify its organic constituents. When these FTIR spectra are compared to the standard spectra of the lubricant, the presence of any hydrocarbon contamination in the lubricant will be apparent.

3. Wash the balls, races, and cage in an organic solvent. Note any significant or unusual wear patterns on the cage material. Study the wear pattern on the inner race, outer race, and the balls, first with an optical microscope and then with SEM. Look for any embedded particles and determine their composition with EDXS. During normal wear, the highly polished surface on the races become dull. Because the preload in disk drives is from the bottom, the wear pattern is off-center on the top side. Look for any craters or any signs of corrosion.

CASE STUDY 9.1: Misalignment of Bearing

Problem. A spindle bearing became noisy and experienced premature failure.

Hypotheses. The field engineer suspected two probable causes: (1) excessive heating of the bearing, and (2) contamination of the lubricant with a sealant similar to Locktite. This sealant, used at the junction of the outer race and the housing casting, was suspected of contaminating the lubricant in the bearing.

Analysis. The external surfaces of the bearing appeared to be clean. The bearing was carefully dismantled. The race wear surfaces and balls did not exhibit wear marks or discoloration. The bearing grease was black and very thick.

The lubricant was washed off with freon. Several shiny particles collected at the bottom of the solvent container. Under a low-power optical microscope, these particles were yellow, with a metallic shine. Figure 9.1 shows SEM micrographs of the wear particles. The EDXS analysis showed the presence of copper and zinc, suggesting that they were brass. When compared by EDXS analysis with the brass cage, it was clear that the particles in the lubricant came from the cage.

The bearing cage was studied with an optical microscope and SEM examination. Surfaces of the cage in contact with the balls exhibited considerable wear at many locations. On one side of the cage, the surface showed wear at the outer diameter, and on the other side, the wear was at the inner diameter (see Fig. 9.2). Figure 9.3 is a SEM micrograph of the uneven wear on the cage.

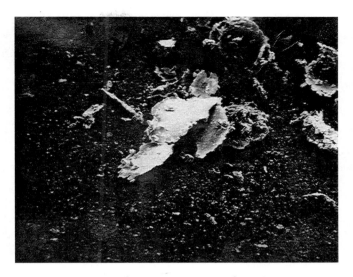

Fig. 9.1 SEM micrograph of metallic particles found in bearing grease. EDXS studies indicate the presence of copper and zinc (brass from the cage) in the particles. 228x.

BEARING CAGE

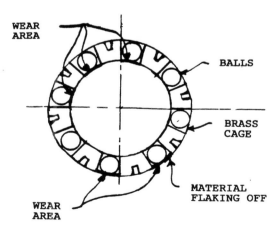

Fig. 9.2 Schematic of the area of the cage where wear was observed.

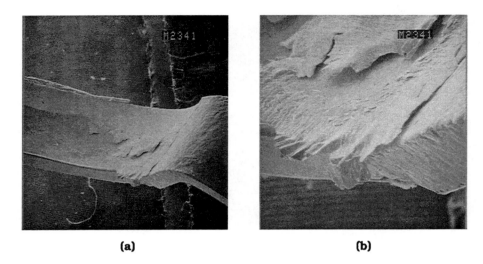

(a)   (b)

Fig. 9.3 Uneven wear on the surface of the brass cage due to rubbing of
the balls. The wear debris flaked off, as shown in (b).

Discussion. Based on its excessive and uneven wear pattern, it
was concluded that the cage experienced wear due to misalignment
of the bearing around the shaft in the drive housing. This misalignment caused an uneven load on the balls (see Fig. 9.4), which caused
uneven wear on the cage. In certain areas, the wear was so high that
the cage fractured (see Fig. 9.5). It should be noted that the brass
cage is significantly softer than the steel balls and races.

The particles generated by wear of the brass cage became entrapped in the lubricant and oxidized into black copper oxide during
rotation of the bearing. The presence of copious amounts of copper
oxide made the lubricant black and thick, rendering it hard to rotate
and noisy.

Because there was no discoloration on the balls and races caused
by oxidation, it was determined that the bearing had not overheated.
It is possible that some heat was generated by the rubbing of the balls
with the cage and the cage debris. However, this heating must have
been quite mild and relatively harmless, because any significant
amount of heating would have discolored the steel balls and races.
The FTIR analysis of the lubricant indicated the presence of only the

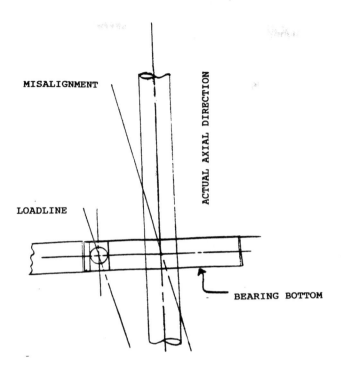

MISALIGNMENT

ACTUAL AXIAL DIRECTION

LOADLINE

BEARING BOTTOM

Fig. 9.4 Schematic depicting how misalignment of the bearing occurs, leading to uneven load.

grease and no Locktite-type sealant. Thus, the hypothesis that lubricant was contaminated with sealant was dismissed.

Solution. Improved process monitoring and control of the assembly of bearings eliminated misalignment-type failures.

CASE STUDY 9.2: Metallic Hard Particle Contamination

Problem. One batch of bearings was found to be noisy after only a few thousand cycles of operation. Several used and a few unused bearings from the lot in question were studied in detail to determine the cause of failure.

Hypothesis. The spindle nose grinding operation introduced particulate contamination into the bearing lubricant and caused the failure.

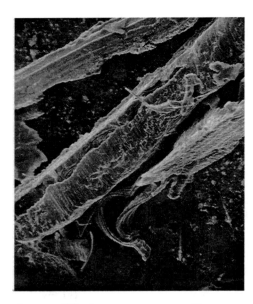

Fig. 9.5 Fractured area of the brass cage caused by severe wear.

Analysis. The bearings were quite clean on the outside. The unused bearings were opened carefully, and metallic chips were found in their lubricant. The lubricant was washed off with freon, which was then allowed to evaporate. The particles remaining at the bottom were found to be magnetic and thus were separated using a permanent magnet. The metallic chips found in the unused bearings were studied by SEM/EDXS. The EDXS compositional analysis indicated that the chips had the same composition as the race material (52100 steel). SEM micrographs of the chips are shown in Fig. 9.6, which clearly shows tool marks on the surface indicating that they were generated by a machining process.

When the failed bearings were opened, the lubricant was darker and was found to contain black particles. The outer race showed abnormal wear and cracking. The black particles were separated from the lubricant and studied with SEM/EDXS. The EDXS compositional analysis indicated that they came from the bearing race material (52100 steel). Typical SEM micrographs of the particles are shown in Fig. 9.7, which indicate that the particles had a relatively smooth surface.

(a)

(b)   **(c)**

Fig. 9.6 Metallic particles, with machining marks that indicate they are machining debris, present in an unused bearing. (a) 300×. (b) 1500×. (c) 1500×.

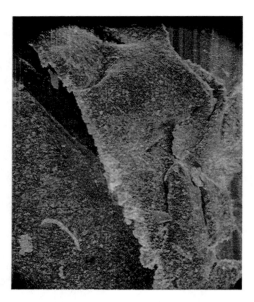

Fig. 9.7 Metallic particles with smooth surfaces present in a failed bearing. Wear of the machining debris during the bearing operation made the surface smooth. 1080×.

Discussion. The unused bearings exhibited bearing material contamination generated by the machining operation used to prepare the smooth race surface. When the bearings with this metallic contamination in the lubricant were used in the drives, the particles of contamination rubbed against the race, producing a smoother surface.

In general, when soft particles are present in the bearing lubricant, they are pulverized and become oxidized. When the oxidized debris is mixed with the lubricant, the lubricity changes, making the bearings hard to rotate. When hard particles, of the same hardness as the bearing material, are mixed with the lubricant, they abrade the race surfaces, thus generating more particulate contamination and making the bearings noisy.

The particulate contamination in the failed bearings abraded the race surface and made it rough, thus causing failure. Because the particulate contamination was found in unused bearings even before assembly of the spindle, the hypothesis that the spindle nose grinding process generated the contamination could be dismissed.

Solution. It was recommended that the bearings vendor improve the cleaning process to eliminate particle contamination before bearing assembly.

CASE STUDY 9.3: Pitting from Electrical Discharge

Problem. A ball bearing experienced intermittent noise and was examined to determine the cause of failure.

Hypothesis. Welding occurred between the balls and the races.

Analysis. The bearing was opened carefully, and a lubricant sample was collected for FTIR analysis. The rest of the lubricant was washed away with freon. The balls and the races were studied in an optical microscope and by SEM examination.

Figure 9.8 shows SEM micrographs of the balls, exhibiting clear evidence of melting at those locations, as shown by the beading of the type 440C stainless steel ball material. Figure 9.8(d) shows that the ball was welded at the top first and then separated at the bottom as it rolled on the raceway.

Figure 9.9 is a SEM micrograph of the inner race pitting. The pits encircle the race intermittently in the direction of the translative movement of the ball. It appears that the material melted around the pits, and the heat-affected zone is clearly visible around them.

The FTIR spectrum of the lubricant did not completely match that of the standard lubricant (see Fig. 9.10). When the electrical resistance of the lubricant in the failed bearing and the standard lubricant were compared, the resistance of the failed lubricant was found to be about two orders of magnitude higher.

Discussion. The heat-affected zones around the pits on the raceways and the presence of beading around the pits and ball spots indicate that heat was generated at certain locations on the balls and the adjoining raceways. The pattern of the pits and spots indicates that there was momentary heat generation, and as the balls moved relative to the race, the molten metal was pulled in the direction of motion. From electrical resistance measurements, it was evident that the lubricant was a poor conductor of electricity.

The failure can be explained as follows. An electrical charge built up between the balls and the races, and the lubricant, being a poor conductor, was unable to remove the charge buildup. When the charge reached a critical level, it discharged, causing some areas on the balls and raceways to melt and weld together momentarily. As the balls moved relative to the raceways, the weld was broken due to the force of rotation. The charge started building up again and the

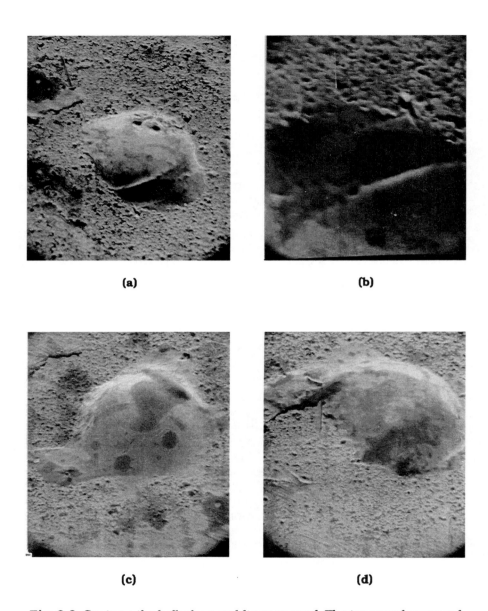

Fig. 9.8 Spots on the ball where welding occurred. The topography around the spots indicates melting of the metal. (a) 610×. (b) 1677.5×. (c) 610×. (d) 610×.

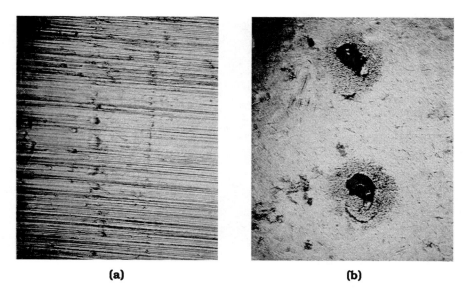

(a)   **(b)**

Fig. 9.9 Spots on the inner race of the ball bearing, where the ball experienced welding. The heat-affected zone around the welded area is clearly visible. (a) 6.1×. (b) 61×.

cycle continued. This process of welding and breaking generated particulate contamination and led to failure. The original hypothesis that actual welding occurred was correct.

Solution. When the poorly conducting lubricant was replaced with a lubricant with good conducting properties, failure due to electrical pitting was eliminated. Periodic measurement of the electrical resistance of the lubricant was undertaken to ensure long life of the bearings.

CASE STUDY 9.4: Preload and Wear Fatigue

Problem. Several bearings from one batch were found to be noisy during routine operation and were analyzed to determine the cause of failure. Several unused bearings were also analyzed with the failed bearings.

Hypothesis. Because the preload was higher than normal in this batch, excessive preload could be a problem.

Analysis. The failed bearings appeared to be clean outside and were opened carefully for optical microscopic examination. The lu-

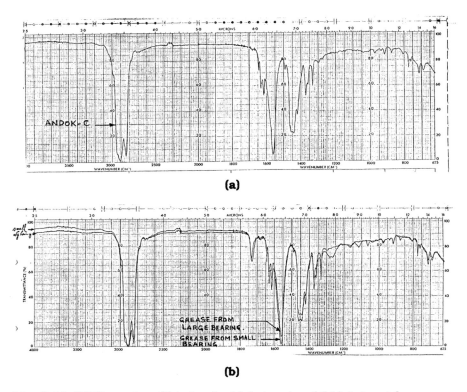

Fig. 9.10 FTIR spectra of (a) standard lubricant and (b) lubricant from a failed bearing. Because they are very similar, it was concluded that there was no sealant contamination in the lubricant of the failed bearing.

bricant, which contained shiny metallic particles, was washed out with a petroleum solvent. The shiny particles were found to be ferromagnetic and were picked up with a magnet.

SEM micrographs of the particles (see Fig. 9.11) showed that they had smooth surfaces. EDXS analysis indicated a composition similar to the type 440C stainless steel of which the bearings were made. The particles did not appear to be the result of external contamination.

Figure 9.12 shows SEM micrographs of the outer raceway, which reveals smearing, true brinelling marks, and evidence of flaking off-center on one side of the raceway in the direction in which the preload was applied. There was similar, although less severe, damage on the inner race.

Fig. 9.11 Particles that flaked off the outer race surface. SEM, 67×.

Fig. 9.12 Surface of the outer race, showing that material flaked off because it was stressed over its fatigue limit. SEM, 134×.

Fig. 9.13 Typical linear actuator.

The unused bearings were opened and studied. There was no particulate contamination in the lubricant.

Discussion. This appears to be a case of rolling contact fatigue failure. The fact that the damage was concentrated off-center in the preload direction indicates that failure was due to excessive preload. Because the new and unused bearings contained no particulate contamination, it could be deduced that the particles found in the lubricant were generated by flaking from the raceways.

Solution. The preload was decreased by about 50%, and the potential for failure was eliminated.

9.5 Linear Actuators

In a hard disk drive, to access information stored at a particular location on a disk, the head needs to move to the specified track location. Because high access speed is one of the most desirable characteristics of a hard disk drive, a carriage with head arms is attached to an actuator that can move very fast.

There are two different types of actuators, linear and rotary. Linear actuators are convenient for larger drives, for example, 356- and 229-mm (14- and 9-in.) disk drives. Rotary actuators are widely used in smaller drives, such as 203-, 133- , 89- , 64-mm (8-, 5.25-, 3.5-, and 2.5-in.) disk drives, primarily because of their compactness in smaller form factors, faster response times, and convenience. Even in a smaller drive, if there is significant distance between read and write elements, e.g., as in dual-element magnetoresistive read and inductive write heads, linear actuators are preferred.

In a linear actuator, the heads move in a straight line "in and out." The carriage with the head arms has ball bearings that move on a hardened flat surface called a "rail." The rail is generally made of hardened stainless steel or an aluminum alloy that is anodized or coated with a wear-resistant layer, e.g., a plasma-deposited coating. Figure 9.13 illustrates a typical linear actuator.

Two distinct modes of failure are common in the linear actuator bearing-rail combination. The first type is failure of the inside of the bearing, which is very similar to the type of failure experienced in the spindle bearings discussed above. The second type of failure occurs in a manner that is unique to the linear actuator, i.e., failure of the interface between the bearings and the rail. This mode of failure is discussed in detail below.

9.5.1 *Materials*

Bearing surfaces such as the outer surface of the outer race and the rail, should have sufficiently high hardness to provide acceptable wear resistance. For the bearing, hardened martensitic steel (alloy 52100) is suitable. Because corrosion and consequent particle generation cannot be allowed in disk drives, type 440C stainless steel, which is much more corrosion resistant than alloy 52100, especially in humid environments, is a more appropriate choice of material.

For the rail, either type 440C stainless steel or an aluminum alloy with a hard coat is used. Anodizing provides a hard, wear-resistant coating several micrometers thick. Because the top layers of an anodized coating are generally porous and weak and are potential particle generators, the top 10% of the coating thickness is lapped off to obtain an optimum coating. Passivated type 440C stainless steel appears to be a suitable material for rail construction, but tribologically, two identical materials rubbing against each other pose other problems.

Type 440C stainless steel is a suitable material for both the bearing and the rail when the components are considered individually. However, it is not a good choice when they are considered as a tribological unit. In a tribological unit analysis, both mating parts should be analyzed simultaneously to obtain an accurate representation of the working environment and factors affecting performance.

What are the disadvantages of two stainless steel parts rubbing against each other?

First, when the mating parts in a tribological unit are identical, they tend to adhere to each other. The energy of adhesion between the two is very high, which leads to more wear due to the mechanism of adhesive wear.

Second, when the mating parts are identical, they have similar hardness values. If a hard and abrasive contaminant particle comes to the interface between them, the mates do not yield to each other. As a result, abrasive wear continues, generating more and more particles. But if one of the mates is significantly softer than the other, the abrasive contaminant can be trapped in the softer matrix.

Third, stainless steels favor the polymerization of stray organic molecules, with the help of frictional heat, to form "frictional polymers" on their surface. These polymers significantly increase the friction between the two mating parts and thereby increase the wear at their interface.

Materials wear by the following primary modes: adhesive, abrasive, corrosive, and surface fatigue (Ref 5).

In the case of adhesive wear, one surface attracts and pulls away the atoms from its mating surface. Slowly the surfaces become rougher with the generation of particulate contamination, which creates more wear. The adhesive wear increases as the energy of adhesion (discussed below) between the mates increases. This type of wear is more or less predictable because it can be estimated from the materials properties of the mates.

Abrasive wear, (also discussed below) is caused by the presence of hard and abrasive particles consisting either of adhesive wear debris or contaminant particles. It is quite unpredictable and is the most deleterious of all the wear mechanisms.

In the case of corrosive wear, one or both of the components undergo corrosion in a corrosive environment. The relative motion between the mates removes the passive corrosion products, thus exposing the reactive material to more corrosion.

In surface fatigue wear, the repeated loading and unloading cycles caused by the relative motion of the mates fatigue the surfaces and create surface and subsurface cracks, leading to fracture and particle generation.

9.5.1.1 Energy of Adhesion

Suppose $r(a)$ and $r(b)$ are the surface energies of two sliding materials and $r(ab)$ is the interface energy. Then, per Ref 5, the energy of adhesion, $W(ab)$, is given by:

$$W(ab) = r(a) + r(b) - r(ab)$$

The wear coefficient, k, is found to be proportional to the friction coefficient, f, which is proportional to $W(ab)$. If the sliding parts are identical, $r(ab) = 0$. For similar materials, $r(ab)$ is quite low. In cases of $r(ab)$ being close to zero, $W(ab)$ is high, making f and k very high. When the sliding parts are made of incompatible materials, wear will be low.

9.5.1.2 Abrasive Wear

Two-body abrasive wear occurs when a hard part digs into a very soft mating part and removes the soft material. Three-body abrasive wear occurs when a stray or contaminant particle comes between the mating surfaces and abrades the material. When the mating parts are relatively close in hardness and have smooth surfaces, three-body abrasive wear predominates.

Three-body abrasive wear can be controlled by making one of the mating parts slightly softer than the other, in which case the hard contaminant particle is pushed into the softer mate during sliding and will be captured there so as not to cause any more damage.

When two identical materials with identical hardnesses slide over each other, neither of the mates yields to the contaminant, allowing it to remain at the interface and cause more surface degradation.

9.5.1.3 Frictional Polymers

A mixture of hydrocarbons deposited at a sliding interface could be cross-linked into a higher molecular weight material by frictional heat and pressure generated by the sliding components. Because the products of such cross-linking have properties similar to mixed polymers, they are called frictional polymers. Exposure to hydrocarbons and organic vapors at concentration levels of several parts per million can lead to the formation of frictional polymers (Ref 6).

Although palladium and platinum are known to produce many frictional polymers, metals such as chromium, molybdenum, and gold are capable of producing at least moderate amounts of them.

Frictional polymers tend to form at the stainless steel-stainless steel interface and adversely affect tribological properties. In a disk drive, there are many sources of airborne organic contaminants, such as adhesives, plastic components, and lubricant in actuator bearings. These organics could lead to frictional polymer formation.

Despite these drawbacks, the use of stainless steel for both bearings and rails is still popular because of the ease of surface preparation and process control it offers. Also, some of the problems associated with the identical bearings and rail combination can be improved by the application of lubricant at the interface and by improving the cleanliness of the drive environment.

9.5.2 *Failure Modes*

Some of the common failure modes of the linear actuator bearing-rail interface are discussed below. Either inaccuracy in the multiple bearing system or rail misalignment, or both, can cause misalignment between the bearings and the rail, which causes the bearings to dig into the rail on one edge, generating particulate contamination and causing off-track error and a loss of signal due to a change in the head flying height.

Inadequate surface finish or improper passivation of the rail could cause failure. Rail surface preparation is the most critical step in ensuring a good interface. A smooth surface finish, with surface roughness less than 0.05 μm (2 μin.) is required to minimize wear and particle generation. Under-passivation decreases wear resistance, while over-passivation makes the surface brittle and generates loose flakes (per the case study below).

If the bearings contain excessive lubricant, it can leak out onto the rail and interfere with the smooth running of the bearings over the rail.

Contamination at the interface is a common source of failure. Particulate contamination can be embedded in the surface of the bearings at some stage in their manufacture, e.g., honing, or it can come from the disk drive environment. Any hydrocarbon contamination can lead to the formation of frictional polymers at the interface.

CASE STUDY 9.5: Over-Passivation of Rails

Problem. Signal degradation occurred in a disk drive that was then returned for failure analysis.

Hypothesis. There was either loss of material from the disk or accumulated contamination on the read/write head.

Fig. 9.14 Accumulated debris on the composite head covered its core, causing signal degradation.

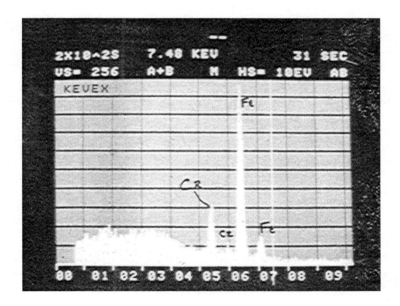

Fig. 9.15 EDXS spectrum of debris accumulated on the head showing the presence of iron and chromium, indicating that a 400 series stainless steel part was generating loose contamination.

Fig. 9.16 Wear marks and debris on a linear actuator rail.

Analysis. The disk drive was opened carefully, and the heads and disks were studied through an optical microscope. Material accumulation observed on the air bearing surface and on the sides of the head (see Fig. 9.14) consisted of shiny particles and organic debris. Elemental analysis (see Fig. 9.15) of the particles showed the presence of iron and chromium in the ratio of approximately 4:1. Because the disk contains iron oxide media, part of this iron signal could have been coming from the media debris. However, the presence of iron and chromium in the ratio of 4:1 in the particles suggested a 400 series stainless steel.

The disk was scanned with an epoxy-coated magnet to collect any particles such as 400 series stainless steel. The particles collected were analyzed by SEM/EDXS and found to contain iron and chromium in a ratio of approximately 4:1. The presence of numerous 400 series stainless steel particles on the disk and head showed that a component of this material was generating particles in the disk drive environment and ruled out the possibility that a stray stainless steel particle was present on the head prior to assembly of the drive.

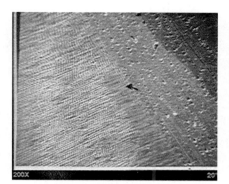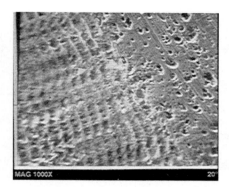

Fig. 9.17 Porosity on the virgin portion of the rail and wear marks on the over-passivated rail showing particle generation due to the movement of the bearing on the brittle surface of the rail.

Examination of various components in the drive showed that the bearings accumulated some particulate contamination, and the rail on which they moved exhibited significant wear and an accumulation of metallic debris (see Fig. 9.16). EDXS analysis indicated that the rail and the debris were 400 series stainless steel. Figure 9.17 shows the virgin area on the rail, which is porous, brittle, and over-passivated. The area of the rail on which the bearings ran showed of wear and particle generation (see Fig. 9.17).

Discussion. The virgin rail area looked over-passivated, and the worn area was much smoother. The over-passivated rail was porous and brittle. When the bearings ran over it, its brittle surface layers fractured and flaked off, generating airborne contamination that accumulated on the disks and heads, causing signal degradation.

Solution. The rail was passivated until a uniform and smooth film was formed. The passivation process (i.e., the concentration of nitric acid in solution, temperature of the passivating bath, and the time of exposure) were optimized to prevent over-passivation.

9.6 Rotary Actuators

The present trend in the disk drive industry is toward smaller drives and smaller form factors (i.e., the arranging of more disks in a smaller package). Rotary actuators are preferred over linear actua-

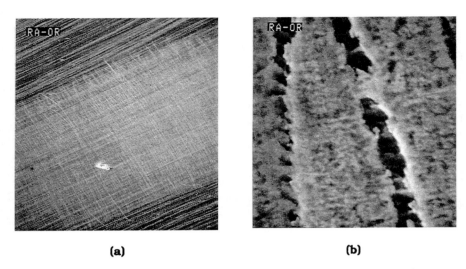

(a)   (b)

Fig. 9.18 Bearing race showing metal-to-metal contact and material removal due to fatiguing due to jittering and restricted movement of balls in a rotary actuator. (b) Higher magnification of (a).

tors in these applications because the former are more compact and convenient and have fast access times.

In a linear actuator, the bearing rotates through 360° many times. The rotary actuator bearing, however, does not rotate through a full revolution, but oscillates through only about 21°. During track-seeking by the head, the bearing vibrates or "jitters," which tends to push the lubricant away from the ball-raceway interface, causing metal-to-metal contact and eventual failure. Figure 9.18(a) shows the type of damage caused on the rails by jittering, and Fig. 9.18(b) is a high-magnification micrograph showing metal removal caused by fatigue resulting from metal-to-metal contact. The metal debris abrades the races and balls, and the cycle continues until the bearing becomes hard to rotate and eventually fails (Ref 7).

Techniques to extend the life of rotary actuator bearings involve ensuring that the lubricant is not pushed away from the ball-raceway interface, by using balls with as small a diameter as possible and a low-migrating lubricant, and by increasing the hardness of the bearing material to increase its fatigue life.

Precision assembly of the bearings on the rotary actuator is one of the most important problems to overcome in improving disk drive performance. One possible solution is use of an integral bearing assembly (Ref 7), in which the shaft is used as the "inner ring" to eliminate tolerance buildup and constricted rotation of the balls.

9.7 Materials Testing

The following parameters should be monitored to ensure the manufacture of a quality bearing:

1. Hardness of races, balls, and actuating rail
2. Purity of the lubricant without particulate and hydrocarbon contamination
3. Use of an optimum amount of lubricant
4. Particulate contamination embedded in races and balls during manufacturing
5. Use of a cage material that is not chemically reactive with the lubricant
6. Surface roughness and optimum passivation of the rail, in linear actuators
7. Preload size, which should be kept as small as possible
8. Prevention of outgassing and particle generation

References

1. *Metals Handbook*, Vol 11, 9th ed., American Society for Metals, 1986
2. J.G. Hannoosh, Ceramic Bearings Enter the Mainstream, *Des. News*, Nov 1988
3. J.F. Dill and R.A. Harmon, Rolling-Element Bearing Technology: Sizing up the Japanese, *Mech. Eng.*, Dec 1987, p 37-39
4. K.D. Mackenzie, Why Ball Bearings Fail, *Automation*, Dec 1975
5. E. Rabinowicz, *Friction and Wear of Materials*, John Wiley & Sons, 1965
6. E. Rabinowicz and S. Webber, The Formation of Frictional Polymers on Noble Metal Surfaces, *Proceedings of the 10th International Conference on Electric Contact Phenomena*, 1980, p 98-101
7. W. Johnson, "Integral Spindle Provides Greater Reliability in Winchester Disk Drive Rotary Arm Actuators," *The SKF Tribologist Newsletter*, SKF Industries, Inc., 1985

Index

A

AA. See *Atomic absorption spectroscopy*
Abrasive cleaning, 114
Acicularity, 40
Acid etching, 90
Acrylics, 197
Activation energy, 53
AC welding, weld qualification, 116-117
Adhesion
    electroless nickel, 80
    EN coating on Mg alloys, 102-103
Adhesion tests, 80-81, 89
Adhesive bonding, 149-154
    adhesive curing, 151-154
    advantages, 150
    contamination removal with gaseous plasma, 151
    materials bonded, 150
    potential outgassing, 154
    precautions for ensuring successful bonding, 150
    surface preparation, 150-151
Adhesive curing, 151-154
    undercured copper coil assembly, 152-154
Adhesives, 12, 17
    thermal stability of, 12, 13, 16, 17
AEM. See *Analytical electron microscopy*
AES. See *Auger electron microscopy*
AFM. See *Atomic force microscopy*
Air bearing, head accumulated contamination on surface, 263-264
Alkali etching, 90
All-ceramic bearings, 306
Alsimag, 240
    ceramic slider material, 30
Alternating electromotive force (emf), 277
Alumina
    addition to magnetic paint to improve disk wear resistance, 29
    addition to silicon nitride ball bearings, 307
    aqueous cleaning and, 214, 215
    chipped, handling damage prior to or during assembly of disk drive, 265, 266
    chipping of disk drive as source of failure, 268-269
    coating of aluminum disks, 43
    disadvantages in disk use, 279
    dispersion and distribution, 41-42, 51
    dispersion seen by transmission electron microscopy, 47
    etching of disks, 44-45, 46
    hard dust particle scratches, 264-265
    incomplete removal of oxide with soldering, 133-134

C

Eutectic tin-lead 63-37 alloy solder, 124-127
Exothermic reaction, 12, 15
Expansion coefficient, 16, 17
Expansion/contraction, 12

F

Failure analysis
    causes of failures, 4-7
    challenges, 8-9
    failure prevention as essence of, 9-10
    methodology, 7-8
    methods used to improve flex circuit reliability, 197-205
    modern definition, 1-2
    traditional concept of, 1
Fatigue
    plug and cable assembly, 181-183
    stainless steel retaining rod, heat treatment effect, 190-191
Ferrite core, 35
Field emission electron guns, 32
Filiform corrosion, contaminants causing, 68
Filiform corrosion under an electrostatic coating, 66-68
Film-like contamination, 208, 209-210
    permalloy plating imperfections, 301, 303
Flash welding, 105
Flex circuits
    aggressive fluxing of copper, mechanical fracture, 173-175
    failure analysis methods to improve reliability, 197-205
    surface roughness effect on strength of copper traces, 167
Flex leads, soldering, 124
Flying height of read/write head, 88
Foam swabs, recommended over cotton swabs, 235, 236
Fotoceram, encasing ferrite cores, 30
Fourier transform infrared spectroscopy (FTIR), 22, 27-28, 211
    aluminum disk human contamination, 227
    ball bearing failures, 309-310, 312-313, 317
    grease bearing of linear actuator, 218
    nitric acid attack on aluminum studied, 227, 228-229
    to monitor coating lubricant supply, 280
Fracture stress, 166
Frictional heat, 138
Friction welding, 105
FTIR. See *Fourier transform infrared spectroscopy*

G

Galvanic corrosion, 4-5, 122, 124
Gamma iron oxide, 37
    coating of aluminum disks, 43
    coating of aluminum substrates, 55

H

Hysteresis loop squareness, 40

I

O

Pull strength, 14-115, 116
Pull testing, 114-115
Pump oils, 150

Q

Quality control, 273

R

Rails, 305
    over-passivation, 326-329
RBS. See *Rutherford back-scattering spectroscopy*
Read voltage, 305
Read/write heads, 207-208
    composite, 207, 240
    handling damage from plastic boxes, 266-268
    monolithic, 207, 240
    pad materials, 242-243
    plating, 283
    solvent stains during testing, 269-270
    technology used in making, 240
    thin-film, 30, 207, 240
    thin-film, aqueous cleaning problems, 215
    thin-film, defect growth on plated media, 260-263
    thin-film, external contamination, 258
    thin-film, handling contamination due to cosmetics, 233, 234-235
    thin-film, hard dust particle scratches on alumina, 264-265
    thin-film heads with silicon on permalloy, 218-219
    Winchester drives, 241
Reducing atmosphere, to prevent decarburization effect on hardness, 169
Remanent magnetization, 277
Resin application process, 163
Resin/filler ratio, 12, 13
Resin/magnetic particle ratio, 17
Resin system, cure advancement of, 12
Resin systems of particulate media, 17
Resistance welding, 105-111
    austenitic stainless steels, evaluation of welding parameters, 114-116
    underheating and subsequent reworking, 111-114
    weld failure caused by edge roughness, 118-122
    weld qualification, 116-118
Resolution, 42
Retaining rod, heat treatment effect on fatigue failure, 190-191
Rotary actuator bearings, 305
Rotary actuators, 322, 329-331
Rubber, thermal stability of, 13
Rutherford back-scattering spectroscopy (RBS), 22
RVA machines, 276-277

S

Salt bath nitriding, 65
SAM. See *Scanning acoustic microscopy, Scanning Auger microscopy*
Samarium cobalt for disk drives, 209
Saturation magnetization, 277
Scanning acoustic microscopy (SAM), 22, 25
Scanning Auger microscopy (SAM), disks with spit mark contamination, 250-251
Scanning electron imaging (SEI) mode, 30-31
Scanning electron microscopy (SEM), 8, 22, 23, 24
    adhesive bond studies, 151
    aluminum disk human contamination, 225, 226
    brake drum electroless nickel plating examined, 83
    bubbles and pinholes caused by substrate defects, 71-72
    bubbling of powder coating caused by moisture studied, 70
    compositional difference vs. brightness of lead-tin solder studied, 127-129
    connector pin mechanical fracture studied, 171
    contamination of stainless steel studied, 222, 223
    copper strands of ground cable, mechanical fracture studied, 178-180
    defect determination on disks, 253-255, 257
    diffusion joints with ultrasonic welding studied, 143
    disk fiber contamination leading to flight interference, 287
    disk magnetic properties, 278, 279
    disks with spit mark contamination, 245, 246, 247, 248, 249
    electroless nickel coating on magnesium alloys, 99
    electropolished stainless steel screws studied, 157, 158
    etching disks for analyses, 44-47
    examination of weld failure caused by edge roughness, 118-121
    flex circuit reliability studied, 198-199
    flex leads for soldering studied, 124-128
    fracture analysis of soldered wire, 160-164
    handling contamination of copper plates studied, 222
    handling contamination of thin-film heads due to cosmetics, 233-235
    hydrocarbon condensation for copper shunts studied, 220-221
    hydrogen embrittlement during zinc plating examined, 168-169
    improper materials selection for screws studied, 172, 173
    incomplete removal of oxide with soldering, 133-134, 136, 137, 138
    insufficient stirring of the solder pot studied, 129-120, 133, 134
    lock screws, fracture caused by improper materials selection studied, 194
    moisture entrapment by intermetallics studied, 218, 219
    motor shaft surface scratching studied, 188-189
    nickel-zinc ferrite pad sintering problems, 270-271
    particle generation from a static ground spring, 272
    passivation of aluminum and soldering studied, 136, 137
    permalloy plating imperfections, 292, 294, 296, 303
    plating and surface porosity, fluid entrapment, and blisters, 288-290
    plug and cable assembly, fatigue studied, 182
    resistance welding of austenitic stainless steels, 106, 107
    rigid disks analyzed, 43-47
    spring failure caused by a sharp bend studied, 183-186
    stainless steel, scribing of, 236, 237
    surface porosity and lubrication study, 279, 280

T

W

Wash weight, 279-280
Water, as source of contamination, 150-151
Water jet cleaning, 214
Wavelength dispersive x-ray spectroscopy (WDXS), 22, 27
WDXS. See *Wavelength dispersive x-ray spectroscopy*
Welding, 105-122
Weld zone, 107, 108, 115
Winchester drives, 14, 30
    disk manufacturing specifications, 274
    heads, 207
    overcoat, 81
    read/write heads, 241
    surface porosity and lubrication, 279
Wire embrittlement, 162
Wrought aluminum alloys for disks, 17
    thermal expansion coefficients, 17

X

XPS. See *X-ray photoelectron spectroscopy*
X-ray computed tomography (CT) scanners, 306-307
X-ray diffraction (XRD), 22, 26
    particle generation from a static ground spring, 272
    voice coil machine mineral deposits study, 224
X-ray fluorescence spectroscopy (XRF), 22
X-ray photoelectron spectroscopy (identical to ESCA) (XPS), 22, 27
XRD. See *X-ray diffraction*
XRF. See *X-ray fluorescence spectroscopy*

Y

Young's modulus, 165, 275
Yttria, as addition to silicon nitride ball bearings, 307

Z

Zinc
    deposits found as a result of improper cleaning with soldering, 132, 135
    manganese-zinc ferrite core in head, 35
    plating and hydrogen embrittlement with mechanical fracture, 168-169
    plating on high-carbon steel retainer clip and mechanical fracture, 167-168
Zinc alloys, electroless nickel alloys, 87
Zinc-aluminum 95-5 solder, 124
Zinc plate, replaced by electroless nickel, 83
Zirconia, debris from ball milling constituents, 46